T0209758

Dr. Angela Longo's

QUANTUM WAVE LIVING WORKBOOK

Tools for Discovering and Living My Eternalself

Balboa Press books may be ordered through booksellers or by contacting:

Balboa Press
A Division of Hay House
1663 Liberty Drive
Bloomington, IN 47403
www.balboapress.com
1 (877) 407-4847

ISBN: 978-1-9822-2124-9 (sc)
ISBN: 978-1-9822-2125-6 (e)

Print information available on the last page.

Balboa Press rev. date: 02/18/2019

WHAT PEOPLE ARE SAYING ABOUT
Dr. Angela Longo's
QUANTUM WAVE LIVING

I met Angela Longo in Thailand few years ago. I was really curious about the work she was doing because I could see people around me literally changed overnight after having a session with her. I could tell that just by looking at them, they felt different.

After my first session with her, I went into a deep transformation myself. My perception was shifting and at the same moment my life and the things that happened to me changed. Even though, I had no knowledge whatsoever about quantum physics (in the first place).
Afterwards, I asked her more details about the way she worked and I ended up being a student of her.

Angela is full of passion and experiences. She always listens and knows how to help you out. She focuses on the root of issues and that changes everything.

Today I'm a quantum life coach and I use her techniques daily. I can say that the techniques work on everyone.

There is this moment in your life when you say : "There will be a before and after that experience."
Well ... Angela and her work is one of these moments.
I will never be able to thank her enough for what she brought into my life.

For the people who want to experience something that will change their perception and beliefs about life, I highly recommend to look into her work and if possible book yourself a session to see how it goes for you.
If you are ready to let go of certain beliefs, I promise you that you will change the way you look at life and, weirdly enough, the way life looks at you will change at the same moment.

Be honest, responsible and ready! You are about to jump into the quantum wave of living your life! (And there is no turning back)

Sebastian Roland
November 7,2018
Perpignan, South France
sebastienrolland6@gmail.com

I had been in the hospital for 10 days diagnosed with Dengue Fever. Two months later nothing was helping.

I am a Licensed Acupuncturist and have been in practice 25 years as well as being a Naturopathic Physician and RN. When I entered my session with Dr. Angela we both commented on there being a lot of heat and wind present in my pulses and distinguished which ones were deficient and which were excess. As we progressed through the Quantum Coaching session I began to feel changes in my body and a relaxation in my psyche/spirit that was very surprising to me since it seemed like we were just identifying some beliefs and finding ways to shift them.

At the end of the session we both rechecked the pulses and much to my amazement there was a distinct change in all of the pulse positions; no needles, no substances ingested, just the Quantum process. It still leaves me very impressed and interested in learning more about this system.

I am still experiencing the benefits and changes from our session together days later, so thanks for that.

Thanks again Angela,

Marian S.
August 4, 2013
(She remained well for months after, when I checked. Dr Angela)

Working with my dearest Dr. Angela Longo in re-writing my original story: such a path of deep opening and discovery of my fullness! Those child-like tantrums around abandonment; my mother issues; waiting for the other out there to fulfill me; waiting, waiting, waiting; the giving-up on me… after so many years of inner work, today, by taking a quantum leap, I see myself, I value myself, I commit to myself, I am there for myself, I do exactly what I desire and that is being my eternal authority of self. Transforming my old loner body genes, empowering the eternal authority of myself, celebrating and uplifting my whole self. What a ride! Angela Longo, you're a dance of life! Thank you for sharing your wisdom with the world!

Natalia Mendez, Counselor
Spanish mother of two teens
Living in Melbourne, Australia

Dear Angela,

I think of you with love and great respect.

You helped me through a very hard time in my life.
Even though we were on a different continent you where there for me and I could easily reach you, no matter when.

You worked with me and you found my inner strength and well being my very personal old me. And then we build that up again to come to the surface and bloom again.

You found me these sentences which I learned like mantras and I kept in my heart and mind, ready to enroll them whenever I do need them again.

Thank you for your help and your efficient work!
I would not be like this after a year.

Love you **Kathrine H.**
Germany
Nov. 28, 2018

August 30, 2015

Dear Dr. Angela Longo,

Hi, warm greeting from Pakua clinic, Tao Garden, 19 September 2015. I would like to say "thank you so much-Khob khun mak krub" for your helpful session, mirror calls – quantum resplendency of relationshifting along with TCM healing called "moxa" which are really wonderful miracle, my knee got stronger Chi within a few times. You help me know more about my quantum wave pattern. Now I know who I am, what are my true desires, and how I can improve myself, transforming limiting patterns to what I really want.

If I want to succeed in my true desires or my wishes, just do it and attend to my true desires to frame a universal self-authority...a delight to know me!

Made the Chi be with you.

Mr. Kridsadapong, Eddy
Hoistic consultant and certified UHT instructor.
In Pakua polyclinic, Tao-garden, Chiang Mai, Thailand.

November 2015

Angela's work is truly a paradigm shift that is phenomenal.
I have been working on myself for many years and work as a counselor. These techniques are simple effective and life changing.

Words do not do it justice. After I take the training with her I will be implementing them into my practice.

Melissa Essence
Australia, Adelaide

QUANTASIA:

Thanks Dr. Angela

The quantum session we did in August was a flying introduction into how fast things can shift. I arrived back home feeling 10 years younger -a few people said that so, it's not just me. My attitude and life situation in work and in relationship seem to have shifted 180 degrees and, surprisingly I feel optimistic about moving these internal shifts into the world. I have been paying attention to dreams and mirror calls too and looking forward to progressing some more quantum leaps.

Tom C. Sydney, Australia
Sept. 20, 2016
Continuing...
Stress lifted....work easy....
Since I did my sessions with Angela i noticed a terrific shift. Every month for the past year I have had intensely stressful and high task load weeks at the end of month-pleased to say that this past week was a stroll in comparison- everything, including the usual unexpected hiccoughs seemed to go quite smoothly, and without the many extra long hours I have been used to doing. Thanks!
Tom Christensen
Sydney, Australia

October 1, 2016

Hello Angela,

I have begun to use the things you e-mailed and they make me yawn like crazy. That usually means things are moving.

The good things and shifts were:
- wonderful weather when I arrived and so almost no jet lag, which is usually a great problem;
- I generally sleep well, has also been a huge problem. A great shift;
- we had a lovely party to celebrate the finishing of a meditation hut;
- my relationship with my husband is easier. He listens more. Important shift;
- I teach tai chi and it works out very well. I have more energy;
-Although I have several health issues I seem to solve them faster

Lous W.
Dutch middle aged woman
Belgium living
Oct 1, 2016

"It's wonderful how dreams bring out our connection between our eternal self and our personal self's mirror-calls."

Faith May, Quantum Life Coach
July 18, 2018
Jackson, Colorado and
Phoenix, Arizona

From Tao Garden, Thailand:
January, 2017
The self-nurturing has been amazing today, □□ □ GC

Dear Angela,
I/you have helped me to see, feel and live quantum reality and personally experience the meaning and power of entanglement and the following emergent miracle.
In my session with you I experienced a headache come and go when replacing the informotion that it was made up of. For me headaches used to be a problem, once they appear they grow. Now my new understanding is able to shift them.
I am living my radiant glow (resplendence) and others benefit and express that to me even at dinner tables in restaurants...i

G C, Physiotherapist,
Perth, Western Australia

February 9, 2017

Dear Angela, thank you for your brilliant teaching. It's been so enlivening to watch a fellow human being realise their fully resplendent illuminated being and I thank you for sharing your wisdom with me so I can help others.

Wow 3 years since I was stranded in Bali and met the amazing Angela Longo who taught me the quantum way ... your teachings have supported me big time this week!! so grateful for this transformational technique ♡ when the student is ready the teacher appears ♡ Quantum Yin

Melissa Essence
November 6, 2018

I am in gratitude to Angela Longo as a fearless grounded angel with deep understanding and a compassionate heart.

She came into my life in 2008 at a very pivotal time.
Her care, healings, herbs and teachings have supported me awakening.
I now use some of her teachings to support many awakening globally.
I am eternally grateful that this humble woman saw my eternal being, the gifts I bring to the world and empowered me to see also.

Much aloha,

Janine Seymour
Near Auckland, New Zealand
Nov.18, 2018

NEWSPAPER EDITORIAL in *Messenger*, Topanga, California

You are here: Home / Columns / Living Well—
The Quantum Life
Living Well—The Quantum Life
January 15, 2015 - By Sage Knight

I awake to the usual sound of roosters crowing. Though my blinds are drawn, I can tell it is still dark out. Snug under the light covers of my lovely bed in a Lanna-style Thai bungalow, I explore my inner climate: calm, mostly sunny, with no clouds on the horizon. I feel wonderful. I begin checking in with core relationships: mom, dad, kids, sweetheart. All good. Then I have a thought: I have no expectations of any of them. They are all free.

On my first trip to Thailand, in July 2015, I met Dr. Angela Longo, a feisty Italian woman with long dark hair and brightly colored clothes. She told me that Thailand has a color for each day of the week, but she was the only one I saw who dressed accordingly. We met at Tao Garden Health Center and Resort. After she'd helped me to decipher my complimentary Chinese astrology chart, we spoke briefly about her book, "Relationshifting," an exploration of applying quantum physics principles to shift belief patterns. At the time, I didn't give it much thought.

When I returned to Chiang Mai that September, Angela and I became friends. I liked her spunk, generosity and childlike nature. I also had the opportunity to experience her quantum coaching: Each time she taught a workshop she used me as her students' test client.

First we shifted the beliefs attached to what she called my "personal selves," of which there are two; next we shifted beliefs attached to my "eternal selves," another two; then we worked with my "resplendent self," that glorious, radiant me which is the communion of all of my "selves" expressing in unlimited unique ways. I will not describe here the entire coaching process, nor the battle with liver cancer which motivated Angela to develop her work. What I will share is a sliver of my own quantum transformation.

My main benefit from working with Angela is not what I gained; it's what I lost. After about a month of doing the work, I stopped experiencing that most debilitating of emotions: shame. This felt profound. I told her I was curious about whether or not the changes would "stick" once I got home to Los Angeles.

"We'll see," I said.

"No, you will not 'see'! It's up to you!" Angela shot back, in her always passionate voice. "It's a mirror-call! Face and embrace your thought that your transformation is not reliable, and then shift it!"
So I did a quantum "BATHWAVE." The process is deceptively simple: I ran my hands over my entire body, saying, "I face and embrace that I don't trust my transformation. I don't trust that what I do lasts. I think I am unreliable, that my efforts are in vain. This thought is painful and I am ready to shift it now." The secret to BATHWAVE-ing is to face and embrace the pattern without judgment,

running fingers over the body with tenderness and a loving touch as you would with a child who needed to stop lying so she could see her true self.

Once I felt complete with that part, I "poured in" the new belief, tracing one hand from my mid-back, up over my head and down my torso and the other hand from my mid-back and down my sacrum, both hands meeting below the crotch. While doing this, I said, "My transformation can, will and is consistent and reliable. I can, will and am consistent and reliable."

I repeated the present tense of that statement for twenty-one days.

Sounds simple, yet, during sessions focusing on each of my different "selves" and the unconscious beliefs associated with them, I found many patterns that did not serve me. We recognized, then faced, embraced and replaced them all, one by one.

By the time I wake up on that lovely, rooster-soundtracked morning, I've lost shame. I am communicating with a level of honesty, confidence and calm I've never experienced before. And all of my relationships feel lighter.

How does this work?

Scientists have discovered that when you separate two electrons, even by vast distances, an action performed on one is immediately reflected in the other, e.g., spin one electron clockwise and the other will spin counterclockwise—at the same time. The altering of the "first" electron does not "cause" a change in the other. They happen simultaneously, like a mirror. This action is called "entanglement," and it appears electrons don't own the rights to it, as Dr. Angela's work demonstrates.

I remain at Tao Garden for two months, and have the benefit of multiple sessions with Dr. Angela. Many of the issues we clear have to do with relationships. One day at breakfast, I hit on a big one, a longstanding availability issue I've had with one of my core peeps back home. Angela shows me it has nothing to do with "other." She points out what my belief is and, as we eat our scrambled eggs, we shift it. From that day forward, this "unavailable" man begins calling me internationally for an hour a day. Here's the wild part: he does not notice the change in his own behavior—even after I point it out! When I shift, he shifts, mirroring my transformed belief in real time.

I've been back at Top O'Topanga for two weeks. Everyone is commenting on how I've changed and with surprise at our present level of relating. My son, who would barely return a text, spent three glorious hours on a hike with me and my Golden Retriever companion, Shiloh. My kids' dad and I shared a table at Jerry's Deli, the first time in five years we've done anything together, and I'm heading to Pennsylvania to meet Mr. Now-Available's new grandniece.

Most of the time, I'm happy to simply be home with Shi. Without all that extraneous stuff in my head, I like being alone, nurturing my most important relationships: with self (all of them!), the Earth, my breath, and, of course, Shi, my always resplendent Golden Retriever!

Resources: To contact Angela: angelalongo.com. "Relationshifting," the first edition book: http://bookstore.iuniverse.com/Products/SKU-000559969/Relationshifting-Tools-for-Living-Quantum-Resplendency.aspx. New edition "Quantum Wave Living".

Sage Knight is a Shamanic Writer and Mentor. She and Shi live in Topanga, CA and welcome your visits to www.SageKnight.com.

Contents

WHAT PEOPLE ARE SAYING ABOUT v
INTRODUCTION xxvi
RESPLENDENT EMBODIMENT xxvii
THERE ARE ONLY TRANSFORMATIVE CONVERSATIONS CALLED MIRROR-CALLS xxviii
IT IS ALL LOVE LETTERS xxix
DISTRACTING MYSELF xxx
QUANTUM "WAVES ARE THE UNIVERSE: WAVES ARE THE MEANS OF COMMUNICATION"* xxxi
INTERACTIONS GIVE ME NEW OPTIONS FOR EXCHANGE EVEN MY SMART MACHINES ARE IN ENTANGLEMENT WITH MY WHOLE SELF.. xxxii
VISION IT FORWARD! SUMMARY OF QUANTUM WAVE LIVING xxxiv
UNDERSTANDING QUANTUM WAVE LIVING xxxiv
I AM EMERGENT MIRACLES REACHING THE EDGE OF THE UNIVERSE xxxiv
I DO NOT NEED TO KEEP THE SAME BODY CONDITIONS xxxv
ACCESS ALL THE INFORMATION OF THE UNIVERSE: EMERGENT ENTANGLEMENT IS THE STUFF OF MIRACLES xxxv
WHAT IS MY AUTHENTIC QUANTUM WAVE PATTERN? xxxvi
THE FOUNDATION OF THE UNIVERSE IS THE INTERACTION OF "EXCHANGE"; THAT IS 'FOREVER LIVING' xxxvii
DECADE 1 WHAT IS THIS QUANTUM WAVE LIVING? 1
DAY 1 THE QUANTUM WAVE OF LIVING 1
The first secret of the quantum world: 3
The second secret of the quantum world: 3
The third secret of the quantum world: 3
DAY 2 YOU EXTEND OUT AS A QUANTUM WAVE SELF FILLING THE UNIVERSE 6
OUR QUANTUM WAVE NATURE MIGHT BE IMPORTANT 7
WHAT KIND OF INFORMATION MAKES UP MY PATTERNS? 8
THANK YOU GANDHI FOR YOUR RESPONSE, IN AN EMAIL 9
MORE FUN THAN NUMB 11
DAY 3 INTRODUCTION TO EIGHT BEHAVIOR LOVE LETTERS I GIVE MYSELF, AWAKENING MY RESPLENDENCY 14
ACCUSING/JUDGING 15

BATHWAVE TECHNIQUE: (This motion is done while simultaneously saying the statement of the BATHWAVE #1 below out loud or whispering in order to access all five senses.) 16
DAY 4 BLAMING 19
DAY 5 COMPLAINING, CRITICIZING, AND COMPARING 20
DAY 6 LYING 21
DAY 7 HIDING 21
DAY 8 DENYING 22
DAY 9 DEFENDING 23
DAY 10 JUSTIFYING 23
DECADE 2 READING AND NAMING MY QUANTUM WAVE-LIKE PATTERN CALLED MY QUANTUMCODE 27
DAY 1 MAKING A LIST OF 6 THINGS I DID THAT I WAS PROUD OF WHEN I DID IT 28
DAY 2 SPECIFICALLY, HOW DID I DO MY FIRST ACCOMPLISHMENT? 29
DAY 3 REARRANGING ORDER OF QUANTUMCODE 32
Day 4 ADD NEW WORDS ONLY WHEN YOU FEEL GUIDED 32
DAY 5 INSTANT QUANTUM LEAP HIGH-QUEUE 33
DAY 6 DESCRIBING MY ABIDINGSELF 34
DAY 7 CHECKING YOUR ABIDINGSELF CHOICE 35
DAY 8 NAMING MY ABIDINGSELF WHICH WILL BE MY NEW SECOND NAME 36
DAY 9 COMMUNION IN ONE-NEST!!! 37
DAY 10 YOUR FOREVER PARTNER 38
DECADE 3 MAGINATION: LIVING MY QUANTUM WAVE PATTERN IS MY MAGIC-NATION: 40
"I AM A QUANTUM MOVIE PROJECTOR, NOT A CAMERA; EVERY SCENE IS SELF BEING SEEN." 40
AM I MAKING RERUNS? 41
THE FAMOUS DOUBLE SLIT EXPERIMENT 41
QUANTUM PLASTICITY 42
DAY 1 WHO IS IN CHARGE OF LIVING MY LIFE? 42
DAY 2 VISIONING MY DAY 45
DAY 3 SEEING OTHER'S ABIDINGBEINGS 47
DAY 4 MAGIC-NATION HAIKU 48
DAY 5 "You are in everything around you! So look around" 49
DAY 6 MAGINATION REQUIRES A FOCUS AND AN INTEREST OR A QUESTION 51
DAY 7 The Universe is Good And Benevolent. How? 53
DAY 8 Seeing Options Without Judging 54
DAY 9 "Would I be willing to do or say this eternally and have others do or say this eternally to me?" 55
DAY 10 NOTHING TO FIX 56
OPTIONAL METHOD CALLED GAME to use in DECADE 4 56

FINDING THE ANSWERS TO ANY QUESTION USING 6 AREAS OF LIVING AND/OR 5 ELEMENTS, AS WE DID IN GAME 61
DECADE 4 REPLACING OLD PATTERNS IN RELATIONSHIPS: RELATIONSHIFTING MIRROR-CALL METHOD (M&M's) 64
DAY 1 Relationshifting: REPLACING YOUR STARTER BODY'S PAST 64
DAY 2 KNOT ME? YES YOU, COULDN'T BE! THEN WHO? 67
DAY 3 OUR QUANTUM WAVE PATTERN IS OUR CATALYST 69
DAY 4 TRANSFORMING MY CREATIVE EXPANDING FOCUS... 71
DAY 5 TRANSFORMING MY CREATIVE OPEN SACRED SPACE 72
DAY 6 BEING MY SELF-AUTHORITY 72
DAY 7 AM I MY OWN BEST FRIEND? 74
DAY 8 DO YOU RESPECT YOURSELF? 75
DAY 9 DO I RESPECT ALL THAT I DO? 76
DAY 10 DISEMBODYING RELATIONSHIPS 77
QUANTUM REPLACEMENTS FOR RELATIONSHIFTING 81
DECADE 5 QUANTUM DREAMWAVING LOVE LETTERS 83
DAY 1 DREAMWAVING TECHNIQUE 83
AUDIO RECORD YOUR DREAM and/or WRITE IT DOWN 83
DAY 2 DREAM EXAMPLE 86
DAY 3 DREAMS SPEAK THE QUANTUM LANGUAGE OF VISIONS 88
DAY 4 NOT HAVING FEELINGS RUN MY LIVING 89
DAY 5 QUANTUMCODE IMAGE OF ME 90
DAY 6 CONVERT EVERYTHING I HEAR TO IMAGES 90
DAY 7 NOTICING DOING DREAM OPTIONS WITH FEW WORDS 91
DAY 8 MY LIVING IS A PICTURE OF MY DYING 93
DAY 9 SENOI INDIAN DREAM CULTURE 94
DAY 10 SUMMARY OF DREAMWAVING: MAGINATION IS VISIONING 95
DECADE 6 GETTING TO KNOW ME. GETTING TO KNOW ALL ABOUT MY BODY/ENERGY 100
DAY 1 I AUTO KNOW 100
DAY 2 WISDOM OF THE BODY 101
DAY 3 WHAT THE PAIN...MRI 102
DAY 4 KNEE-D TO KNOW HOW LOW I CAN GO...FLEXIBILITY 105
DAY 5 EMPOWERING MY LIVING 106
DAY 6 DREAM IT..LIVE IT 108
DAY 7 IN MY GIVING IS MY RECEIVING, IN MY RECEIVING IS MY GIVING... 108
DAY 8 FIVE ELEMENTS OR WISDOM OF THE ENERGY 111
DAY 9 FIVE ELEMENTS FAMILIA 112
DAY 10 THREE RELATIONSHIP CYCLES INCLUDED IN FIVE ELEMENTS 114
DECADE 7 Grounding my Earthen Element of Wholeness AS A TRIUNITY 119
ESSENCE OF EARTH: GIFT OF EMBODIMENT 119
DAY 1 QUANTUM WAVE SMILING PRACTICE 123
QUANTUM WAVE SMILING PRACTICE 124

DAY 2 I AM "ANEW" SPIN HAIKU 124
DAY 3 STOMACHING MY WORDS AND ATTITUDES 125
DAY 4 YUMMY TUMMY VALUES AND EMOTIONS 126
DAY 5 BEGIN AGAIN WITH SPLEEN AND PANCREAS PATTERNS 127
DAY 6 OPENING SPLEEN/PANCREAS THOUGHTS AND HABITS 128
DAY 7 SPLEEN/PANCREAS WORDS AND ATTITUDES OF WHOLENESS 129
DAY 8 REPLACING SPLEEN/PANCREAS VALUES AND EMOTION PATTERNS... 130
DAY 9 EMBODYING ME AS I DESIRE TO EARTHEN LIVING 131
DAY 10 MOVE EXPRESSLY ON THE EARTH 132
DECADE 8 METAL/AIR ELEMENT SELF-WORTH 135
I am taking these 10 days to explore, expand with and transfigure my own metal/ air self-worth with my BATHWAVEs! 135
DAY 1 Essence of Metal/Air: Transfiguring and Valuing Myself 135
DAY 2 MOVING ON FROM WHAT NO LONGER SERVES ME 136
DAY 3 LARGE INTESTINE THOUGHTS AND HABITS 137
DAY 4 GLOWING STRENGTH 137
DAY 5 DANCE WITH LIVING 138
DAY 6 FAILURE IS NONEXISTENT 139
DAY 7 WONDER OF THE UNIVERSE 140
DAY 8 LUNG WORDS AND ATTITUDES 140
DAY 9 VALUE AND WORTH 141
DAY 10 TIN MAN'S HEART 142
DECADE 9 ESSENCE OF WATER IS *NOT* A STATE OF EMERGENCY; WATER IS A STATE OF EMERGENT-SEE MIRACLES 145
THIS IS THE SECRET OF WATER 145
QUALITIES OF WATER 146
DAY 1 CLARITY HAIKU 146
DAY 2 WATER CRYSTALLIZES 147
DAY 3 WATER BELIEVES THERE IS NO SUCH THING AS FAILURE 148
QUANTUM SMILING 148
Day 4 KIDNEY IS GENTLENESS 149
DAY 5 BLADDER IS SELF-DIRECTING 149
DAY 6 KIDNEY IS SAFE, SECURE AND CONFIDENT POWER 150
DAY 7 KIDNEY IS THE CLARITY OF EXPRESSION 151
Day 8 MICROCOSMIC ORBIT TO WATER ATTITUDES OF OPENING AND TRUSTING 152
DAY 9 HONEST, GENUINE AND EXPANSIVELY AUTHENTIC 153
DAY 10 MELTING WATER'S ICEBERG OF FEAR 153
DECADE 10 WOOD MOTION RELATING: WOOD I OR WOODN'T I BEE TREE-MEND-US. 157
WOOD IS THE SEED OF SELF-INITIATION 157
DAY 1 GERMINATE AND INITIATE SELF-ORGANIZING SEED = BELIEVE 158
DAY 2 REPLACING OLD WOODY BELEAFING 159

DAY 3 SPROUTING= ACTION 160
DAY 4 THOUGHTS OF WOOD DOINGS 161
DAY 5 POLLINATE HABITS I DESIRE 161
DAY 6 BLOSSOM AND RIPEN MY WORDS 162
DAY 7 EXPANDING ATTITUDES OF THE ENVIRONMENT I DESIRE 163
DAY 8 VALUE AND MEANING OF MY UNIQUE QUANTUMCODE IS MY OFFERING 164
DAY 9 WOOD I LIKE TO BE THE REASON FOR MY ILLUMINATEDBEING 165
DAY 10 HOT WOODY EMOTIONS: AM I DOING TREE-MEND-US? 166
DECADE 11 HEART FIRE MOTION 170
DAY 1 WARMING THE FIRES OF MY HEART 171
DAY 2 BATHWAVE TIME FOR HEART BELIEFS 171
DAY 3 WARMING UP FOR FOR HEART ACTIONS: 172
DAY 4 THINKING WITH MY HEART'S UNIVERSE 173
DAY 5 HEART HABITS 175
DAY 6 WARM HEARTENING WORDS 176
DAY 7 IGNITE MY ATTITUDE OF GRATITUDE 176
DAY 8 GIVING MY VALUES HEAT 177
DAY 9 LOVE LETTER OF GRATITUDE NO MATTER WHAT I DO 177
DAY 10 YOUR ULTIMATE PARTNER AND INTIMATE RELATIONSHIP 178
DECADE 12 SMALL INTESTINE FIRE MOTION 180
DAY 1 QUANTUM CONVERSATION IS EXPANSIVE WITH DECLARATIVE QUESTIONING 180
DAY 2 SMALL INTESTINE BELIEF PATTERNS 182
DAY 3 SMALL INTESTINE ACTION PATTERN REPLACEMENTS 182
DAY 4 SMALL INTESTINE THOUGHT REPLACEMENTS 183
DAY 5 FULLY PRESENT FOR SMALL INTESTINE HABITS 183
DAY 6 USING NURTURING CREATIVE WORDS 184
DAY 7 REPLACING SMALL INTESTINE'S PASSION 184
DAY 8 EVERYTHING NOURISHES ME AS I ADD MY MEANING 185
DAY 9 GIVING MY QUANTUMCODE MEANING TO EVERYTHING I DO TODAY 185
DAY 10 IS EVERYTHING NURTURING ME IN THIS MOMENT? 186
DECADE 13 PERICARDIUM FIRE MOTION DECADE 188
DAY 1 PERICARDIUM MOTION 188
DAY 2 BEGINNING WITH PERICARDIUM BELIEFS 189
DAY 3 BATHWAVE TIME FOR PERICARDIUM ACTIONS 189
DAY 4 BATHWAVE TIME FOR PERICARDIUM THOUGHTS 190
DAY 5 BATHWAVE TIME FOR PERICARDIUM HABITS 191
DAY 6 BATHWAVE TIME FOR PERICARDIUM WORDS 192
DAY 7 BATHWAVE TIME FOR PERICARDIUM ATTITUDES 193
DAY 8 BATHWAVE TIME FOR PERICARDIUM VALUES 193
DAY 9 BATHWAVE TIME FOR PERICARDIUM EMOTIONS 194

DAY 10 EVERYONE IS A RESPLENDENT CONSTITUENT OF MY LIVING 195

DECADE 14 TRIPLE HEATER FIRE MOTION 197

DAY 1 TRIPLE HEATER MOTION IS THAT EVERYTHING IS IN HARMONY WITH ME 197

DAY 2 TRIPLE HEATER BELIEF REPLACING 198

DAY 3 TRIPLE HEATER ACTIONS 198

DAY 4 TRIPLE HEATER THOUGHTS 199

DAY 5 BATHWAVE TIME FOR TRIPLE HEATER HABITS 199

DAY 6 BATHWAVE TIME FOR TRIPLE HEATER WORDS 200

DAY 7 BATHWAVE TIME FOR TRIPLE HEATER ATTITUDES 200

DAY 8 BATHWAVE TIME FOR TRIPLE HEATER VALUES 201

DAY 9 BATHWAVE TIME FOR TRIPLE HEATER EMOTIONS 201

DAY 10 PRACTICE SEEING OTHERS AS ETERNALBEINGS HELPS ME EXPERIENCE MY RESPLENDENTBEING 202

DECADE 15 MELT AND REPLACE FROZEN AGE LIMITS, QUANTUM STYLE QUANTUM TOOLS OF ABUNDANCE 205

DAY 1 MELTING FROZEN AGE PATTERNS 205

DAY 2 FINDING THE ISSUES OF ANY FROZEN AGE USING 6 AREAS OF LIVING AND/OR 5 ELEMENTS AS WE DID IN GAME 206

DAY 3 MELTING AGE-BERGS 208

DAY 4 BRRR... 209

DAY 5 I'M MELTING...A FROZEN AGE ASPECT OF MYSELF 209

DAY 6 REWRITING YOUR PAST IS A QUANTUM GIFT 210

DAY 7 NEW 'UNFROZEN' LIVING 210

DAY 8 IN MY GIVING IS MY RECEIVING, IN MY RECEIVING IS MY GIVING... 210

DAY 9 SOMETIMES MONEY GETS A "BAD" RAP 211

DAY 10 EXCHANGES OF ENERGY MANIFEST WAVES (WOLF, M. (2008)) 212

DECADE 16 MY ORIGINAL STORY IS MY ESSENCE AND INCIPIT* 214

DAY 1 WHAT IS MY ORIGINAL STORY OR ESSENCE 214

DAY 2 EXAMPLE OF INCIPIT DISCOVERED FROM CHILDHOOD DAILY HORRORS MANIFESTING IN BODILY SYMPTOMS 216

DAY 3 YOUR ORIGINAL STORY IS THE MEDIUM CALLED "SPACE" THAT MY QUANTUM WAVE MANIFESTS IN 218

DAY 4 219

DAY 4 AWARENESS OF MY ORIGINAL STORY BENEFITS ALL 219

DAY 5 DISCOVERING MY EMPRISE 220

WHAT IS RESPLENDENT RELATIONSHIP? THE REAL R&R 220

DAY 6 MY RESPLENDENT TEAM 221

DAY 7 RESPLENDENT HUMAN RELATIONSHIP 222

DAY 8 THE NEW ABIDINGBEING IN MY HUMAN RELATIONSHIPS 224

DAY 9 QUANTUM WAVE LIVING IS AN IMPORTANT CONVERSATION BETWEEN MYSELVES THIS MOMENT 225

DAY 10 RESPLENDENT EMBODIMENT IS HELPFUL FOR RELATIONSHIPS AND IS NEW THIS MOMENT 227
GLOSSARY **230**
Glossary 2 by Roger B. Cotting (2018) **237**
Bibliography and Reading List **255**
APPENDIX NOTES ON PHYSICS: **263**
NERD ALERT! **263**
OPTIONAL QUANTUM WAVE THEORY FOR THOSE WHO DESIRE **263**
QUANTUM THINKING VERSUS ONLY BRAIN THINKING 263
WAVES ARE PRODUCED BY ENERGY EXCHANGES 264
STRENGTH OF A SOLID LIKE A CRYSTAL 266
About the Author **267**

Preface
By Kelly Wedin

Have you ever felt that you have patterns in your life that continuously repeat themselves? Do you continue to have disagreements with the same people, have health symptoms that keep coming back, or emotions that continue to be triggered? I did.

I had become quite depressed, felt surrounded by emotional triggers, continued to have ups and downs with my health, was unhappy in my personal life and work life. I finally decided I needed to take a break from it all. That's exactly what I did. I resigned from my job and temporarily moved to Thailand where I was determined to find alternative healing practices that could help me live a happier, healthier, and more enjoyable life.

I was beginning to notice change. I was happier, healthier, and enjoying life. I had incorporated many self healing practices into my lifestyle. A couple weeks before I would be leaving Thailand, I felt there was something missing. I was concerned about the patterns in my life. I knew physical and cultural transitions had been difficult for me in the past and I was afraid I would fall back into the same roots, the same patterns: the patterns that prevented me from living the life I desired. I was looking for more, I was looking for Dr. Angela and her sharing of quantum wave living.

I am now living the life I truly desire and it is so simple, once you learn about the three secrets to quantum wave living.

- *Awareness of my unique and specific homemade pattern of my quantum wave.*
- *Understanding "entanglement" and how our waves overlap in the universe.*
- *Choosing the information we desire to put into action in our lives, results in "informotion" creating emergent miracles each day.*

This book is designed to guide you on your own personal self-transforming journey. Dr. Angela will share her own personal experience with you as she guides you through different stages of quantum wave living.

MY FIRST AWARENESS OF EMERGENT MIRACLES

"One day while working with Dr. Angela, I physically experienced a dull, achy pain on my sternum at the same level of my nipples, for the first time. It appeared suddenly after making some profound and powerful shifts. Dr. Angela mentioned it was the acupuncture point, conception vessel 17 which opens the lungs to easier breathing. The name of the point is ancestral chi. I massaged it for a moment. Dr. Angela asked questions which brought up my fear of my future and not having an abundance of energy, money, and not being in the right environment. I face and embraced my feelings about my future and replaced it with "I am living with abundant energy in the environment I desire, in this moment". My eyes widened as I instantly felt the pain disappear, with the replacement of my new quantum shift. ""Informotion", which is the information we are living is the basis of matter and sensation and that's what you just experienced!", Dr. Angela exclaimed.

During my process I have experienced deep relaxation, a release of tension, and calming sensations. One time after our session, I felt a bit off when I returned home. Through my awareness I was able to identify the emotion I needed to shift. I was able to quickly shift this emotion and felt rejuvenated. This is a process of living that works! It is not an overnight fix.

"QUANTUM TANTRUM"

While writing this preface I encountered my first tantrum at 3 o'clock in the morning and I was "sick and tired" of the physical symptoms I was having. When my tantrum was over, I identified some areas that I needed to face and embrace, I made some quantum replacements, resulting in a calm, relaxed state and was able to return to sleep with less pain. The next day I spoke with Dr. Angela about this tantrum and I identified that "I love pain and suffering". I was shocked, learning this saddened me. I also felt relief knowing I could replace this. I began the BATHWAVEing process by loving my pain and suffering. I replaced this with; "I can, I will, and I do live my pleasure and desires on my own and I love it!". I immediately felt some relief with my physical symptoms.

You may wonder how this process will work? The Quantum Wave of Living Workbook will take you through several different Decades: a Decade is 10 days in length. Each day you will have a morning and evening exercise. It would be great for you to have a notebook to write your reflections, quantumcode, and replacements in. This process truly works, with your commitment.

Two great things about the book; you can do it when and as you wish. With your own work and commitment you will soon experience how simple quantum wave living truly is. You are already doing it most likely without awareness. Dr. Angela will guide you to living the life you truly desire quantum wave style.

Now there are a few things you need to know. Dr. Angela has developed quantum living vocabulary words like: informotion, entinglement, mirror-calls, wavicle, etc.. You will soon learn the meanings of these beautiful and powerful quantum "descriptionary" words. You will find these words described in the glossary at the end of the book. Throughout the book there will be offerings for optional readings from Dr. Angela's book *Relationshifting Tools for Living Quantum Resplendency* and videos on her YouTube channel Dr Angela Longo Quantum Wave of Living. These are great resources when you have time to dive deeper.

Begin on your days off say on a weekend because the forward and first two days of Decade 1 take some thought and time.

Your journey to begin your awareful journey towards self-transformation and resplendency starts now.

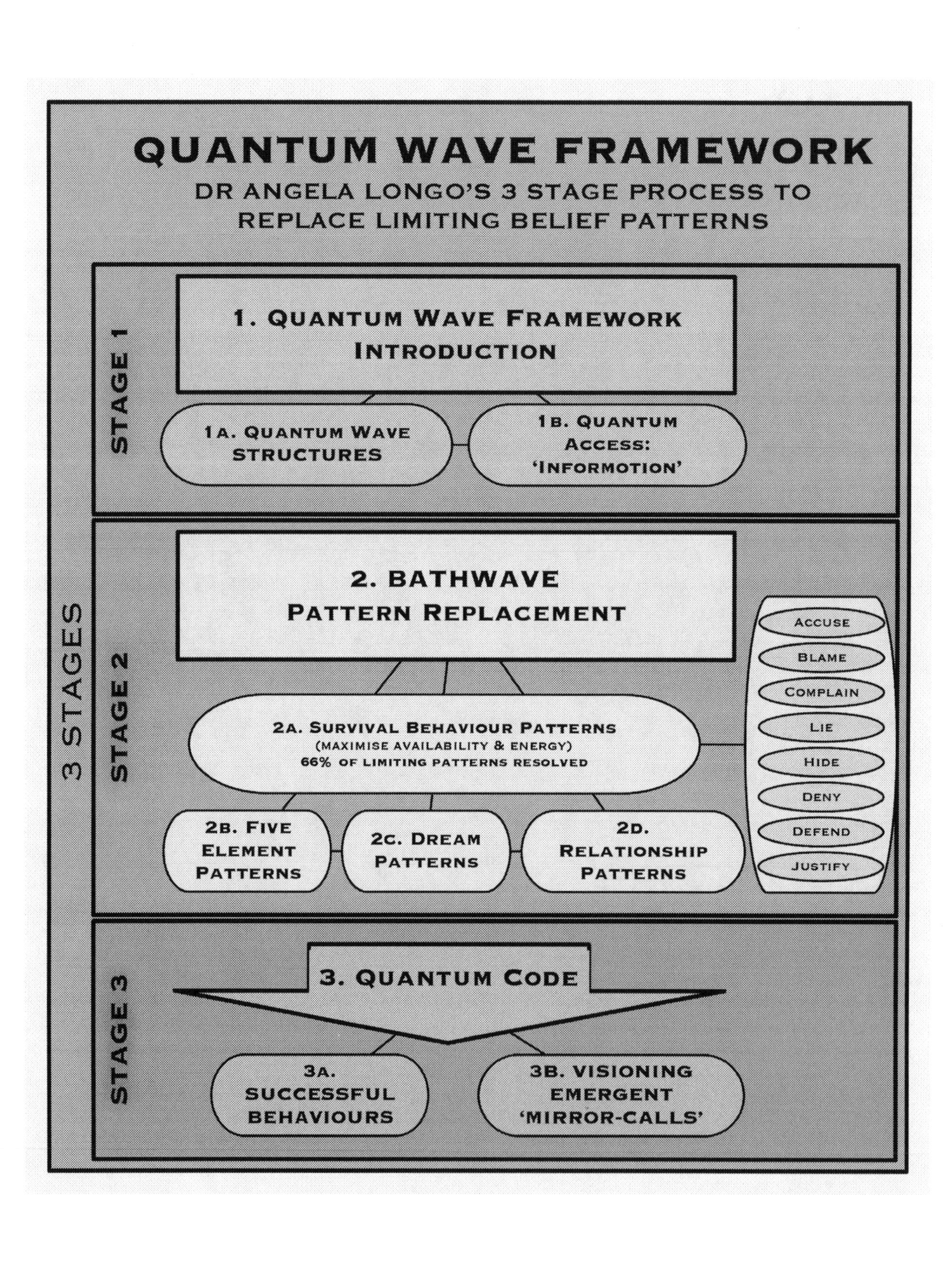

QUANTUM WAVE FRAMEWORK
DR ANGELA LONGO'S 3 STAGE PROCESS TO REPLACE LIMITING BELIEF PATTERNS
3 STAGES
STAGE 1
1. QUANTUM WAVE FRAMEWORK INTRODUCTION
1A. QUANTUM WAVE STRUCTURES
1B. QUANTUM ACCESS: 'INFORMOTION'
STAGE 2
2. BATHWAVE PATTERN REPLACEMENT
2A. SURVIVAL BEHAVIOUR PATTERNS
(MAXIMISE AVAILABILITY & ENERGY)
66% OF LIMITING PATTERNS RESOLVED
ACCUSE
BLAME
COMPLAIN
LIE
HIDE
DENY
DEFEND
JUSTIFY
2B. FIVE ELEMENT PATTERNS
2C. DREAM PATTERNS
2D. RELATIONSHIP PATTERNS
STAGE 3
3. QUANTUM CODE
3A. SUCCESSFUL BEHAVIOURS
3B. VISIONING EMERGENT 'MIRROR-CALLS'

QUANTUM WAVE LIVING FRAMEWORK

Stage 1. Introduction to the Quantum Framework

Have no fear. Though I use a quantum explanation, these steps work without understanding it. I hope you find it interesting and fun.

1A). My vast, universe-size quantum waves structure the body of my life's matters anew as fast as the speed of light. How?

1B). Just add my desirable 'informotion', accessed also through entanglement, for new me.

Stage 2. BATHWAVE technique replaces the past limiting patterns.

2A). First, he eight bottomline survival patterns: accuse, blame, complain, lie, hide, deny, justify and defend are each replaced. Jump into the BATHWAVE, which faces and embraces...loves the limiting pattern... which melts it. Then replace it with my grace... in my core and throughout my universal quantum recording.

2B). Using my five element framework from Traditional Chinese medicine along with the six areas of my life, I discover my whole body of life's limiting patterns to replace with my true desires.

2C). Dreams are sacred teachers in the quantum language of pictures. Daily events can be used the same way. They give us unlimiting options to replace with BATHWAVEs.

2D).Replacing mirror-calls of my personal aspects in relationships is very self- empowering. I may discover a particular frozen age pattern to replace with people.

Stage 3. Reading My Quantumcode-being

3A).. Discovering my unique vast abiding partner using six things I've done that I am proud of and giving them a name is the encode-decode key for self-guidance and living emergent miracles. Emergent is the basic process of the quantum universe.
Counting this new partner, along with the emergent illuminatedbeing, I am actually triunities of beings and selves. This is the quantum basis of the universal sacred flower of life.

3B). Practices to vision, live, & communicate with my quantumcode-being as an aware illuminatedbeing in this moment. I am never alone.

Imagine living without suffering and strife?
Imagine life that's an emergent miracle this moment?!
Imagine my body instantly new in this moment!?
Imagine accessing new information about my true desires from the universe instantly?!
Imagine that I am really a PHYSICAL trio: my first being as my quantum wave is as big as the universe, and spins in its center, appearing as a particle of my second being, my body. This particle disappears the next moment and manifests a new pulsating wavicle, which means a new particle with my invisible, yet physical quantum wave. This wavicle is a third me as a physical awareness of the conversation between the first two of me, as an emergent miracle!

INTRODUCTION

IT TAKES THREE OF ME TO HAVE RESPLENDENT AWARENESS; I AM BEGINNING AGAIN

I am beginning to be aware of this quantum wave self of me who is called my abidingself, since it is always with me. I have personally shaped this quantum wave-like pattern self who is communicating with my personal body/brain personality. I am also aware of these two selves interacting as an emergent third self called my universal illuminatedself.

All three of me live this way of exchanging my new and different desirable informotion, which I access in entinglement with the universe. My ever transforming illuminatedself, is the witness of my other two interacting, personalself and abidingself.

As my illuminatedself exchanges and interacts with my unique quantum wave pattern called my abidingself, I become "conditions of expansion" enlarging my living and the universe. My personalself begins to be aware of these "conditions of expansion". And I begin again, except that, in this moment, I am awareness of my sacred conditions of expansion.* This is awareful beginning again.

I am my own Fortune Cookie

I am a
Resplendent, Illuminated TriUnity

RESPLENDENT EMBODIMENT

This quantum perception of embodiment is so amazingly powerful and so subtle.

I need to respect and honor my body because it is the manifestation of the whole universe as I desire it. This is the quantum point of view.

Outgoing quantum wave as big as the universe bounces back as my incoming quantum wave, together with incoming waves from the universe. The quantum wave spins in its center and it matters!
Next pulsation, the particle disappears as the quantum wave goes out again. a new spinning quantum wave 'particle' replaces it manifesting a new "wavicle". The particle disappears again to go out. This is me! What does it mean for me?

My quantum waves manifests new 'particles' for me as fast as the speed of light. So why worry about what I have done? It doesn't make sense, since it will be new by the time I worry about it?! So I can just do something different? When I am a creature of habit doing the same old, same mold this moment, that IS the real issue.

Helping each other to enjoy something new is really where it is at: do something new...learn something new, even if it's one new word, one new page, one new chapter, one new movement, one new friend, or one new creation. Just do it- don't think about it!

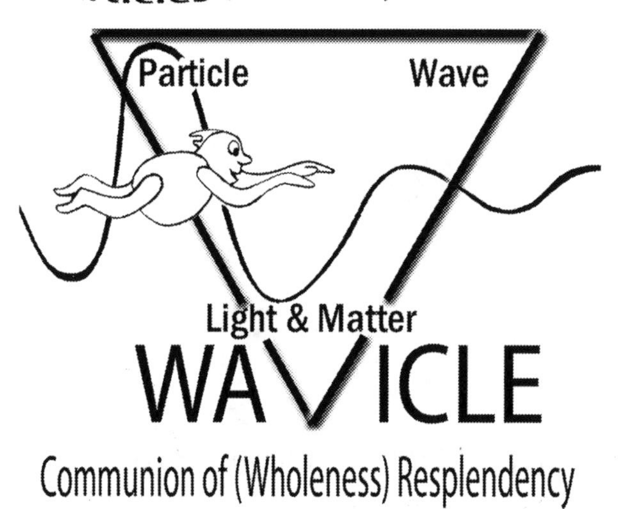

THERE ARE ONLY TRANSFORMATIVE CONVERSATIONS CALLED MIRROR-CALLS

One of the gifts of this quantum wave living workbook is that it opens up a scintillating conversation of personal transforming miracles. This is the creative conversation that goes on within my trios. I have two trios. One is virtual, made up of what I can't see, like my beliefs, thoughts, values, attitudes,.. and the other is seen embodied in my actions, habits, words, emotions, accomplishments and so on.

Exercising my ability to converse is a little like learning a new picture vocabulary. It is much easier than verbal, since I have been seeing pictures my whole life. It makes sense after a while when I relax into it. So play with it and have fun, as I am never alone again.

IT IS ALL LOVE LETTERS

All I really need to do is ask a question and the universe responds via my trio immediately. That's the promise of entanglement: the quantum memory gives me response instantly, in fact, it may respond before I know I ask a question! I call them love letters! I just didn't know how to receive and understand the response. *The love letters are in a universal image language in my body's symptoms, … in my relationships,… in my dreams,… in my daily events,… in my creations,… everywhere,… in this moment,… and in everything… in the whole universe for quantum's sake!*

The reason I call them love letters is that they exist to assist us in living the universe of our true desires without judging ourselves. Recognizing my true desires is one of my favorite gifts I call my quantumcode.

My understanding of quantum wave of living has totally enabled me to love my body. My body is a new manifestation of each moment. I no longer resent my body as something I have to carry around like old past baggage.

Another gift is the understanding of quantum memory, which I call GOOGLEVERSE. Knowing I have complete access to googleverse, whenever I desire, is so refreshing! I am very emotional about this awareness; right now as I am choking up and coughing. I am having many coughing 'orgasms', which we lovingly abbreviate as cocos!! Meaning I am replacing the old emotions in my lungs, that don't serve me anymore, which express as mucus. (lungs are informotion of self value, Decade 8)

So many people are interested in having that house, long life, health & wellbeing, doing exercise, taking expensive supplements and on and on. I believe that this is helping, yet I'm realizing that this new awareness is most important and powerful for me.

I am a Love Letter to Myself

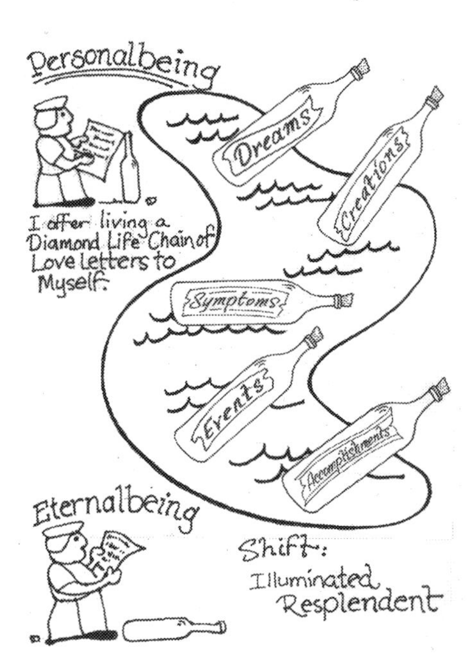

DISTRACTING MYSELF

Anything that takes me away from the enlivening self-sustaining existence of my awareness as my illuminatedbeing in its raw nature is just a distraction for me. So I AM aware when I find myself distracting myself from authenticity. The love letters can get loud. Sometimes they are not what I think I desire.

Like when my dictation function on my computer becomes a mirror-call for me, typing words I may be feeling though I am saying different words. It is even swearing at me. I'm laughing so hard as I try and dictate this chapter. My dictation function has become my good friend and miracle. Mirror-calling me to my resplendence.

I feel now that I may be able to complete my workbook, finally. I'm upset yet so grateful because now I know why it's taken me years to do this. I also understand why I need to write my books in conversation with my editor and/or mentors.

I also realize that I need to go out and talk live around the world. YouTube never captures the alive me. I've studied with Lisa Nichols and now know why the greatest orators speak live around the world. It's the conversation that is important.

Quantum Wave Living Workbook is about how to have conversations mostly with my trios. Yes, I have two trios, my virtual trio selves and my embodied trio beings. I hope it opens up that sacred ability that I see occasionally in my family, friends, and clients.

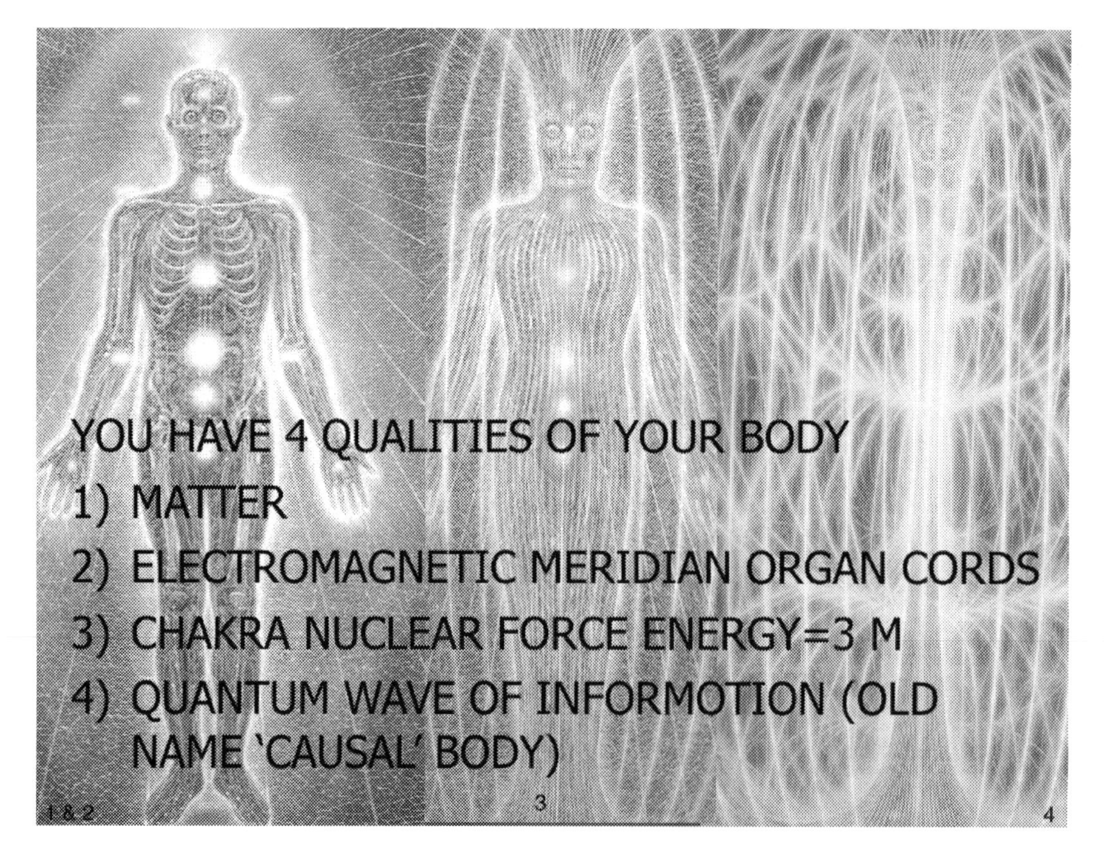

QUANTUM "WAVES ARE THE UNIVERSE: WAVES ARE THE MEANS OF COMMUNICATION"*
Wolff, M. (2008)

I have been realizing that relationship is a principle of quantum wave entanglement. Emergent miracles require interaction with the universe; each moment and each person is a universe! I can interact with people as universes and it creates a whole new creative expansion of living.

What would I say to the universe? On TV or on Oprah...if I knew that other civilizations were listening to what I would say? That really expands my creativity. Like the movie *Arrival*- everything becomes possible. I can value everyone's form of communication, no matter the style since quantum world is the basis of the whole universe.

It is a bit like the expansion of the money system to cryptocurrency. Everything becomes communication even money. Every form of creativity and expression is a form of communication. My work and even the mundane actions of my life become communications to the universe. Because it's not what I do that is important, it's how I do it!

This is what quantum communication is really making me aware of. That every single breath I take communicates to the whole universe. So my life becomes infinitely important and meaningful. What an impact and a relief!

Stressed disappears as a word. Or better yet as I said in my last book: stressed spelled backwards... is desserts! So everything becomes desserts as I wish it.

Everything in the Universe
is made up of Particle/Wave

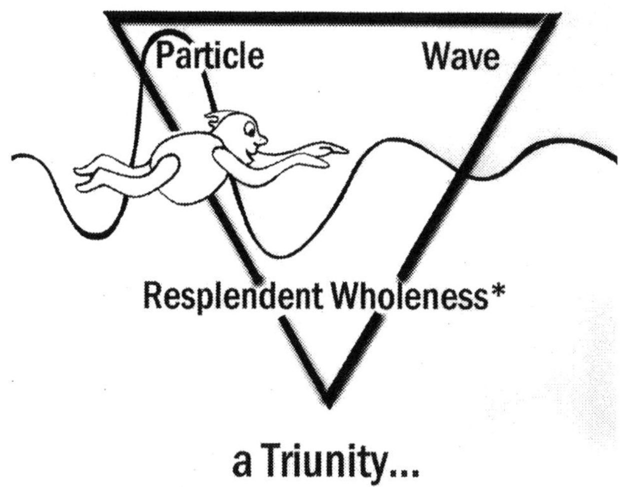

a Triunity...

*electrons, protrons, neutrons, quarks,
bosons and light, which are the building blocks
of atoms, which we are!

INTERACTIONS GIVE ME NEW OPTIONS FOR EXCHANGE EVEN MY SMART MACHINES ARE IN ENTANGLEMENT WITH MY WHOLE SELF

The purpose of all interaction is exchange. Exchange was one of my quantum mentor's, Roger B. Cotting's, favorite words. So there are no mistakes, there are only exchanges. If I do not like what I just did, or what just happened, exchange it.

That's what I have trouble doing...exchanging!? I keep believing that I could lose something; which is never true in the quantum world. I became a packrat which is my Chinese astrology animal: an earthy Minnie Mouse, Disneyland style, and the oriental rat, which is the first animal enlightened Taoist style.

I'm laughing hysterically about that! The famous Italian Pop singer, Lou Monte, who was my mother's favorite had a great song about a mouse, partly in Italian called Pepino, a rascal. Now it makes sense why I love that song. I can feel the real earthy love of my mother, which an Italian never has enough of.

Dancing was the main way I felt embodied when young...and then when I grew older, other adult exercises helped, but I was deathly ill periodically, which means, they were not working at real embodiment. There are good techniques in this workbook to help one with this. My dictation function is reminding me that I am feeling guilty about bothering my editor with my great need for conversation. Lets motion on.

VISION IT FORWARD!
SUMMARY OF QUANTUM WAVE LIVING

UNDERSTANDING QUANTUM WAVE LIVING
I AM EMERGENT MIRACLES REACHING THE EDGE OF THE UNIVERSE

What the quantum wave reveals that makes my living different than what I have been doing for thousands of years are these following amazing understandings.

I am invisible quantum waves as big as the universe, which spin in their center, making the appearance of a 'particle of matter'. They are turning around and going out again as a quantum wave to the edge of the universe… bouncing back! This quantum wave comes in instantly spinning like a new 'particle' and goes out again as a wave, as fast as the speed of light. Add new information into the wave and it is again a new 'particle'.

This is what my personal body is made of.
My so-called 'particles' are really quantum waves spinning, appearing as if a particle and being replaced as fast as the speed of light! Can't get much faster than that!? Then why do I look the same with the same body conditions???

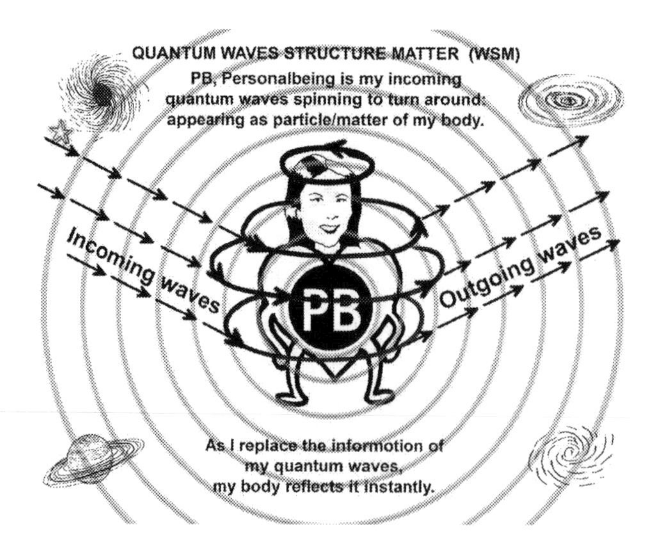

This is the great question whose answer is creative.

I DO NOT NEED TO KEEP THE SAME BODY CONDITIONS

A Naturopath acupuncturist walked into my office in Thailand dying from 2 month old Dengue Fever, in and out of hospitals. Western medication failed. Traditional herbs failed. Three hours later after reading her quantum wave pattern and replacing her old informotion of BATHWAVEs (beliefs, actions, thoughts, habits, words attitudes, values and emotions)...her Dengue was gone. We were both stunned. I kept in touch with her. The Dengue Fever never came back either.

When one works on the fourth 'body' that was misnamed the 'causal' body because it seemed to 'cause' the other three to change instantly, one got miracles or so they were named. I call this 'body' the quantum wave informotion body.

My quantum waves are as big as the universe, which is amazing. And from the experiment called the famous double slit experiment, I learned that I somehow manifest this matter of the universe. I now know that I do this through the 'emergent entanglement' of new informotion I add into my life and therefore, my quantum waves. This IS the stuff of miracles.

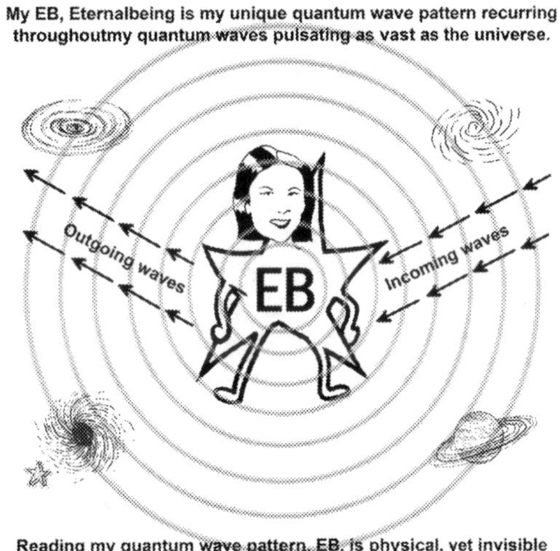

ACCESS ALL THE INFORMATION OF THE UNIVERSE: EMERGENT ENTANGLEMENT IS THE STUFF OF MIRACLES

This is a mouthful, and this is exactly what I'm talking about: 'entanglement' means I have access to all the information of the universe, stuff I have never experienced before. As a biochemist PhD from UC Berkeley, I believe this phenomenon is thanks to my quantum waves of my DNA. There is some research to support this.*

We each have that ability. All I have to do is ask questions. Then understand the picture language of the universe in my dreams, in my relationships, in my bodies, my daily events, and in my creations. Basically this moment, my visions, are speaking of the emergent miracle of my quantum waves.

These quantum waves are me and they are the universe, all in communication. My quantum waves are very individual and unique rendering special messages that I offer my self and everyone, thanks to the interaction with the universe.

I remake myself at this point without cause and effect. I do it by exchanging. **It is my choice to exchange new in-for-motion** for the old information. I do it instantly again by taking in new informotion that I desire and then I trust the emergent quantum process. I can't cause it to happen. That is what quantum wave living teaches me how to do, simply. How? Read on.

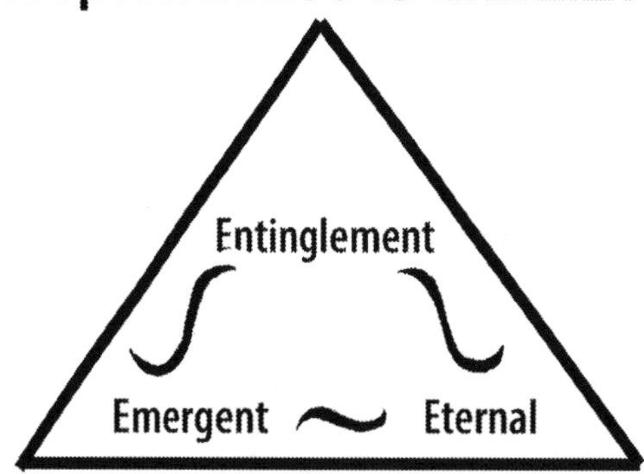

Emergent is New & Different
replacing old ... Expansive.
Quantum equivalence: particle/wave pops
in and out of existence...

Entinglement is Process of Accessing
My Deepest Desires.
Quantum equivalence: one particle/wave process
reflected in the other instantly (beyond cause & effect)

Eternal is Pattern of recurring doings
Quantum equivalence: particle/waving area morphing
recurring pattern...

Zestful

Y (embodifying) YOU

WHAT IS MY AUTHENTIC QUANTUM WAVE PATTERN?

I can read my quantum wave pattern that is authentically, individually, uniquely me. I have the ability by looking back at moments in my life, when I felt proud of what I did, to know exactly what I am doing and have done in those moments. The actions of how I did that, are showing me the pattern of my quantum wave-like doings.

My quantumcode is a pattern that repeats itself moment to moment that I am glad about doing. Or, as I did for so long, I can resist living my pattern, without knowing that I am! These quantum wave living awarenesses and tools helped me transform resistance, and may assist one as one may desire to do this!!

Quantum waves are made up of this informotion that pulsates and undulates throughout 'space/time', my living and the universe. When I look at enough of these moments from my past when I felt proud of what I did, I can then verify my authentic pattern of me pulsating throughout my living. This process is outlined in this new Quantum Wave Living Workbook. I can then exchange this new pattern for my old way of living, and have a new regenerating body of quantum wave living.

THE FOUNDATION OF THE UNIVERSE IS THE INTERACTION OF "EXCHANGE"; THAT IS 'FOREVER LIVING'

And then I can observe the interaction of my selves: my personalself interacting with my quantum wave patternedself, and when I do, I become my true illuminatedself awareness. I can see the miracles that manifest emergently. This means that I can see surprises when the informotion I choose comes together in new ways. This informotion called 'mirror-calls' I am labeling the stuff of miracles.

RIB IS MY NEW WAVICLE
RIB is Emergent of the interaction of my PB (body/brain) and EB (quantum waves).

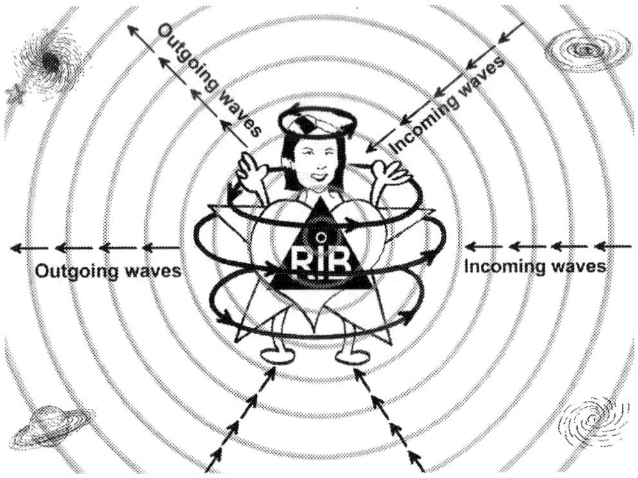

As my PB and EB add new informotion they desire, the body of my life reflects that instantly by entanglement as an emergent miracle.

I am making yet another type of exchange, as this third emergent resplendent self, which is aware and witnesses the interaction of my other two selves: personalself living in communion with my unique quantum wave patternedself. In manifesting, it seems this threeness is required for real awareness throughout my living. One of me witnesses the other two interacting.

Then my resplendent witness interacts with my quantum wave patternedself again and manifests what I call 'conditions of expansion'. These are the sacred conditions or states of my living with resplendent awareness that expand the universe and that manifest living as I truly desire. So *the very foundation of life is one of exchanging constantly.*

When I dare to bare
My heart naked, I turn my
Scared into sacred.

* I thank my patient teachers for this understanding: the late Roger B. Cotting and the honorable Dr. Lam Kong and Master Mark A. Longo

Haiku for HI. Q.

(Haiku for heart intelligence quotient)

When you dare to bare

your heart naked

You turn your scared into sacred

DECADE 1

WHAT IS THIS QUANTUM WAVE LIVING?

DAY 1
THE QUANTUM WAVE OF LIVING

MORNING READING:
Each of us is an illuminatedbeing.

The **quantum world taught me three secrets I needed to live this way**.

I was unaware of them only to find out they are naturally amazing! I want to share them with you... let's begin.

I am a spherical quantum wave as big as the universe. This is pictured in Fig. 1.1. These waves are made up of my personal homemade outgoing quantum wave, which is IN RESONANCE WITH my incoming wave. The outgoing quantum wave hits the edge of the universe and bounces back becoming the incoming wave. As one incoming wave returns it is said to be supported by the universe's quantum waves, to the wave's center, as it spins twice, forming a new high-energy particle (electron) that replaces the old particle (electron) in my body. Then it disappears. This process occurs at least as fast as the speed of light, which is very "quick pulsation". The process continues to replace all of the old particles in my body, over and over and over again. Quantum wave living is quick to regenerate a new body with awareness, as I desire.

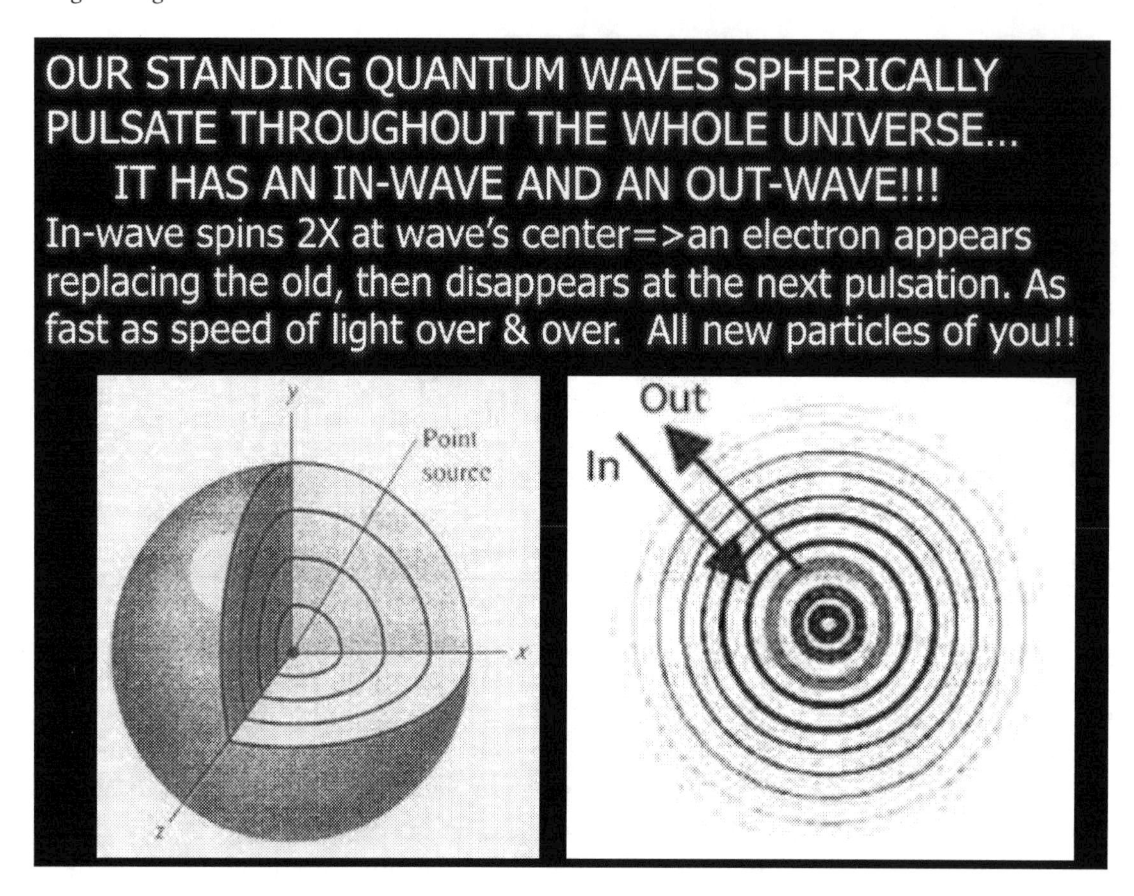

Without self-transforming awareness, my body and living remains the same.

Then why do I keep my diseases, troubles, conflicts? Can I transform them TO LIVE FREE?

The answer is yes...Use these three secrets of quantum wave living as practiced in this workbook, and transformation occurs day by day, point by point, as needed.

One can do it faster than I did, since I was experimenting to put it together. I had great mentors, but still threw a few tantrums. I will be sharing methods to help with any tantrums. I am also discovering easy ways to take advantage of these three secrets of the quantum wave of living throughout this workbook.

The first secret of the quantum world:

I AM REALLY QUANTUM WAVES AS BIG AS THE UNIVERSE.
I am aware that I have a unique and specific homemade eternal pattern of my quantum wave, which reveals my true desires.

In Decade 2, I apply this secret. I make a list of six specific things I have done which I felt proud of accomplishing. These can be small, yet specific events that incorporate how I felt about each event.

The second secret of the quantum world:

THROUGH "ENTANGLEMENT" MY QUANTUM WAVE INTERACTS WITH EVERYTHING IN THE UNIVERSE, CONNECTING AND ACCESSING NEW INFORMOTION INSTANTLY.

In Decade 4, the workbook will guide me through a process to support any relationships. I will begin recognizing mirroring aspects in a relationship and learn how to transform myself through the relationship. This is the beginning of our "mirror-calls" (miracles) relationship method called M.&M.s.

The third secret of the quantum world:

MY QUANTUM WAVE MANIFESTS EMERGENT MIRACLES AND EXPANDS THE UNIVERSE AS I DESIRE!
HOW? This moment is my emergent miracle arising from my new informotion that I live!

Everyday I find and live new and different information that I truly desire and watch my life change miraculously. I am expanding living in the universe as I manifest these desirable surprise miracles this moment. **Expect to be different.** I am adding to my incoming quantum wave what I truly desire. This quantum wave spins at the wave center, which manifests a new particle according to my new 'informotion' (in-for-motion: motion in form).

It disappears only to occur again as fast as the speed of light, PULSATING over and over.

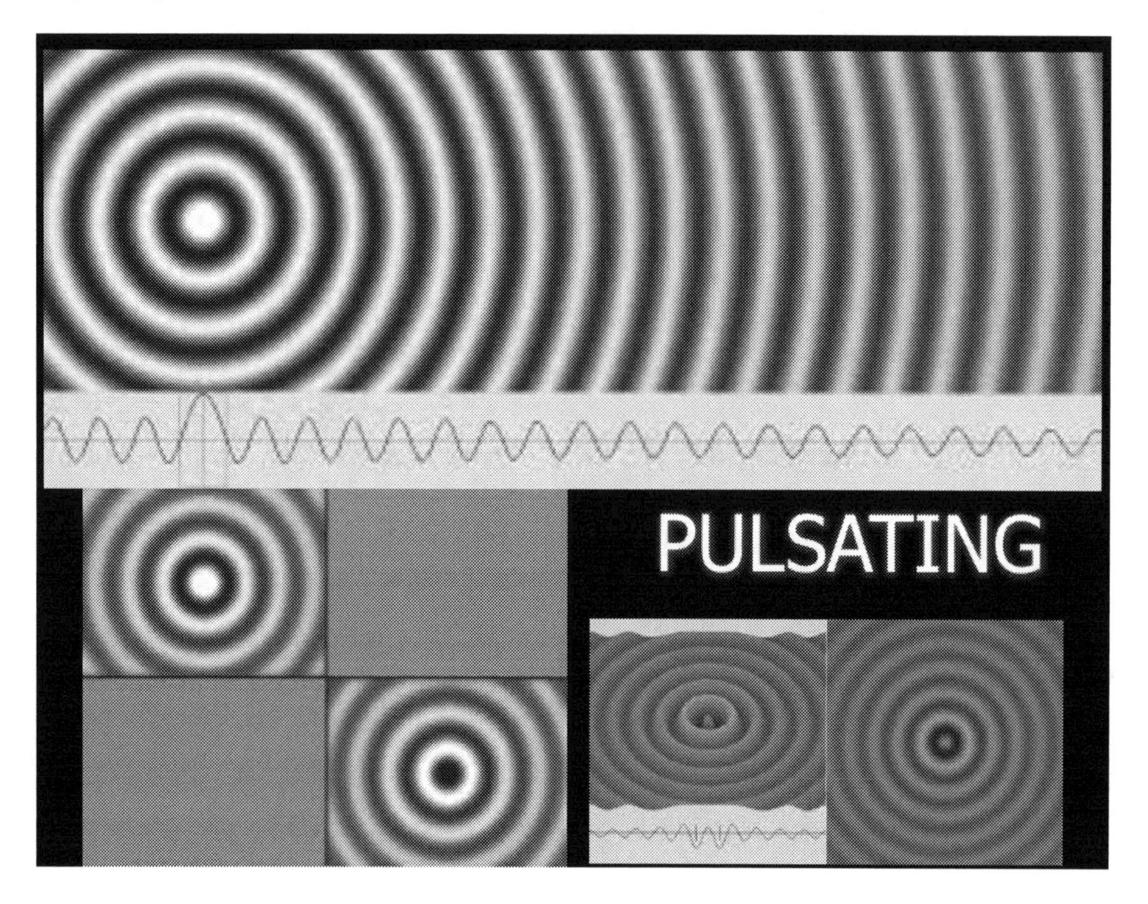

"There is a choice I make every step of the way. It is my re-choice."" Longo, M. (2018) Then the miracle is a surprise. I find the whole process very desirable! Though, I only see my miracles in hindsight. Steve Jobs (2005) gave a commencement speech at Stanford where he also says this very well, "I only see my accomplishments in hindsight."

That's it! Creative miraculous living according to my true desires, enhances wellbeing and transforms relationships without confrontation.

Now I'm putting on my safety belt and enjoying my day, till this evening.

QUANTUM TIP: TURNS OUT:
MY PARTICLES ARE QUANTUM WAVES TURNING AROUND TO RADIATE OUT AGAIN.
WAVES STRUCTURE MATTER!
I MAKE MY QUANTUM WAVES!
WAVES MAKE MY PARTICLES!

Whenever I cannot complete my morning practice, I feel free to do it at lunch or add it to my evening practice.

EVENING READING: What Do I Do with the "Poison" in my Cup?

Twenty years ago when I went to visit a Naturopath with a QXCI machine (an energy diagnostic and treatment apparatus) I was informed of the possibility of liver cancer. At that time, I was an eastern Chinese Medicine Doctor, professor of TCM for 30 years in acupuncture and herbs. With all of this practical knowledge and a western PhD in Biochemistry from UC Berkeley, I had read and applied all of the best research in Western nutrition and positive thinking I had believed in and yet, still all this came about.

"WHAT? ...CANCER?", I was shocked. I thought I was the healthiest person in the world! I went back again two more times, both times receiving a diagnosis of liver cancer and once of lung cancer. I am an energy doctor and I could not ignore this energetic machine.

I sensed that I needed to make a change in my life using something that is more fundamental than even energy. I just didn't know what that was yet. Have you ever felt that way?

I went to one of my mentors, a male elder named Roger B. Cotting. I told him about the possibility of liver cancer.

He told me this story:

"If a student brings a teacher a cup filled with poison and asks, "Can you help me change the poison in my cup, so I can drink from it?"

**The teacher says: "first thing I must tell you is that you cannot throw out the poison that is in the cup.*
You have to do something different.
What you can do is pour fresh water into the cup. You keep pouring and pouring and pouring. The fresh water REPLACES the poison in the cup. So then you can drink from your cup."**

**"What does the fresh water signify in my life?" I asked.
"Living your true desires." he said.
"That should be easy." I answered.**

I went home to start to live my true desires. Looking at my life, examining what I believed to be my true desires, healer, mother, teacher ... I thought I WAS living my true desires. Returning to my mentor I said, "Roger, clearly I don't know what my true desires are." Even at 50 I didn't know my true desires.

Roger B. Cotting smiled and said, "You are correct and now you are ready for the quantum world. The quantum world will show you your true desires."

And that's how this process began...

DAY 2:
YOU EXTEND OUT AS A QUANTUM WAVE SELF FILLING THE UNIVERSE

MORNING PRACTICE:
"There are no particles; just waves whose appearance from a human-scale look like particles" Wolff, M. (2008), p.7.

"The energy of the electron's in-out spherical (quantum) waves... extends out to infinity" Wolff, M. (2008),p.14.

"**Waves are the Universe.**
Waves surround us in our daily life: They are the music and other sounds we hear, the light which conveys the images we see, the heat flow which makes us cold or hot, the radio waves from our cell-phones, and a myriad other sensory events. If we closely and microscopically examine all the matter and phenomena we can, we discover that waves of some sort are involved with all kinds of matter. Are these waves a basic part of the universe? Yes. Because they explain the forces and laws of Nature

and we deduce that waves are the means of communication of all forces, and that waves are the basis of the natural laws. ***These basic waves of the Universe are the quantum waves in space around us.*** If we wish to understand the physical workings of the Universe we need to know the way that waves ... work" (p.88). Wolff, M. (2008)

OUR QUANTUM WAVE NATURE MIGHT BE IMPORTANT

So what IS my quantum wave, which "matters" at its heart, AND WHAT IS IT made up of?
WHAT DOES IT MEAN THAT I AM REALLY A QUANTUM WAVE?!
There are about 10 to the 20th quantum waves in me:
1,000,000,000,000,000,000,000.

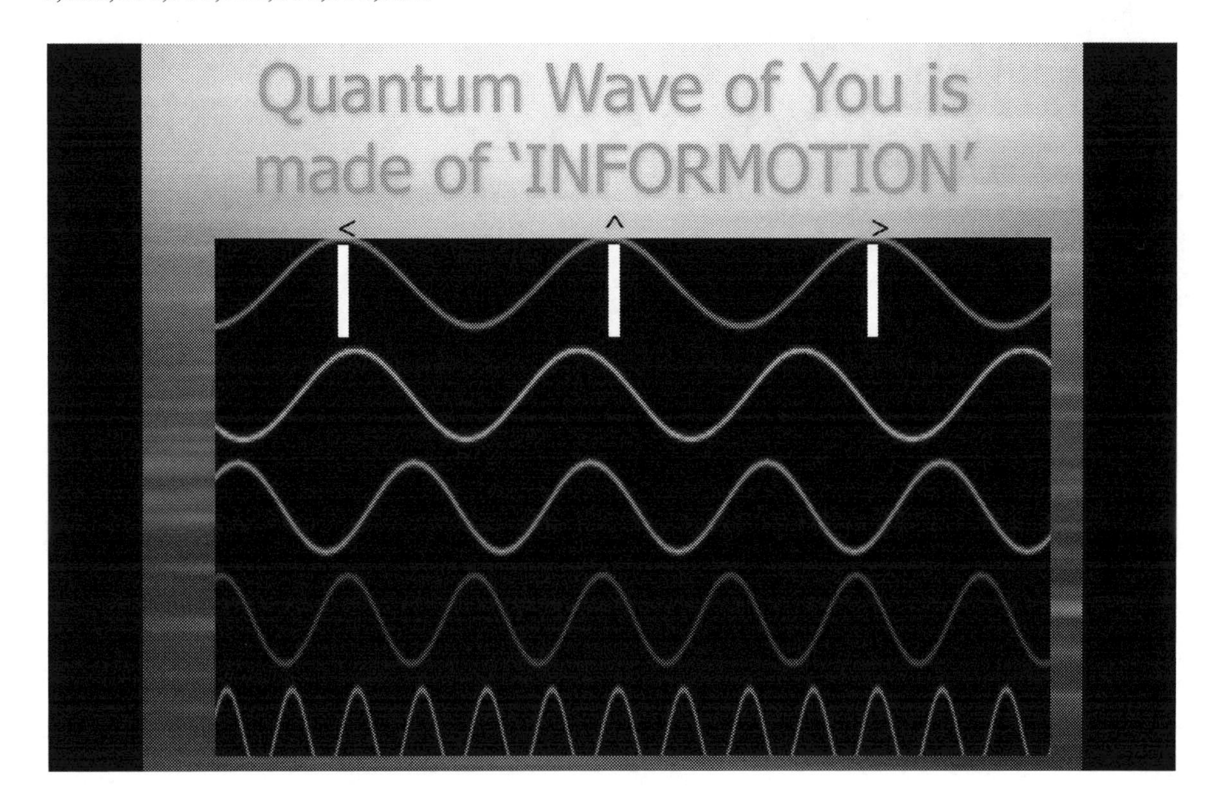

When I slice a quantum wave right down the middle I get a sine wave of peaks and valleys that repeat themselves.
Mark off a complete portion of a peak and valley of a quantum wave and I see the same 'pattern' over and over. And thanks to Schrodinger, his equation expresses this pattern as information. So a quantum wave is made up of a pattern of information that repeats itself over and over radiating out to edge of the universe...Then I am made up of a pattern of information, which repeats itself over and over throughout the universe and throughout my living. Information is made from data. So the universe probably looks something like this according to the quantum understanding.

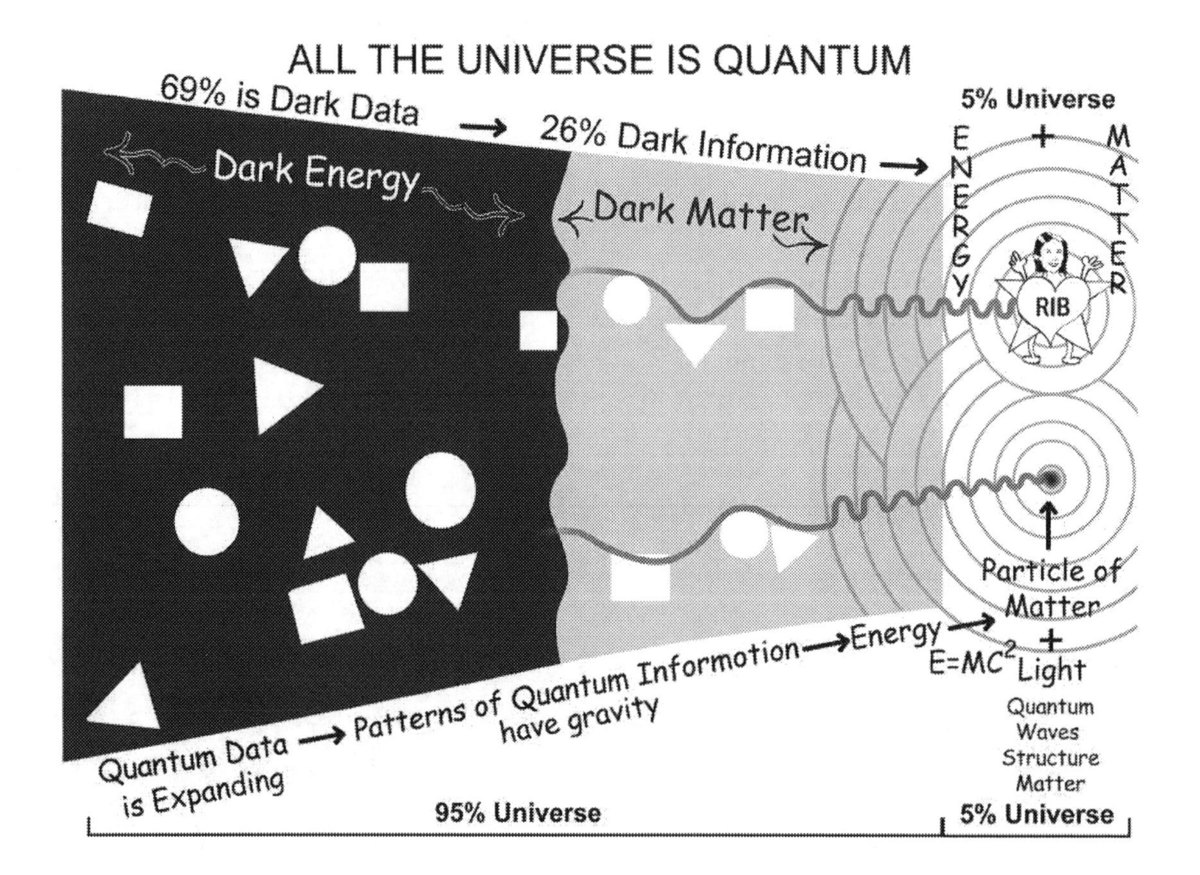

WHAT KIND OF INFORMATION MAKES UP MY PATTERNS?

With all of my training in chemistry, I knew about our body's quantum world of atoms...and I did not like categorizing us as made up of things like information. Chemistry describes that we are only "motion". Our quantum wave patterns are made up of "motion". So I changed the "a" to an "o" and coined a new quantum descriptionary word 'informotion'.

My quantum wave pattern is made up of 'informotion', 'motion' 'in-form'! And that's closer to the worldly-universe.

My quantum wave pattern is made up of motion that informs me ...talks to me ...about me and informs the universe at exactly the same moment.

"INFORMOTION"

- IT'S THE INFORMATION THAT IS THE "REASON" FOR YOUR
- MOTION......DANCE
- YOUR "WAY" & WAVE

WHAT IS BELOW IS ABOVE...!?

So my next question was, "What kind of informotion about me might be contained in my quantum wave that manifests as my particles?".

THANK YOU GANDHI FOR YOUR RESPONSE, IN AN EMAIL....

After asking that question, his quote appeared in an email...
Gandhi's Philosophy for Success is:

"Keep your belief positive because
Your belief becomes your thoughts,
Your thoughts become your words,
Your words become your actions,
Your actions become your habits,
Your habits become your values,
And your values become your destiny."

And your destiny is **YOU.**

YOU is the first letter of the word YOUNIVERSE......!

These words were the "informotions" I was looking for, that were important in my life and would make a difference. This difference would be useful for quantum wave living.

Taking the first letter of all these words and a couple more which my quantum mentor Roger B. Cotting gave me: emotions and attitudes, I got the acronym: **BATHWAVE!**

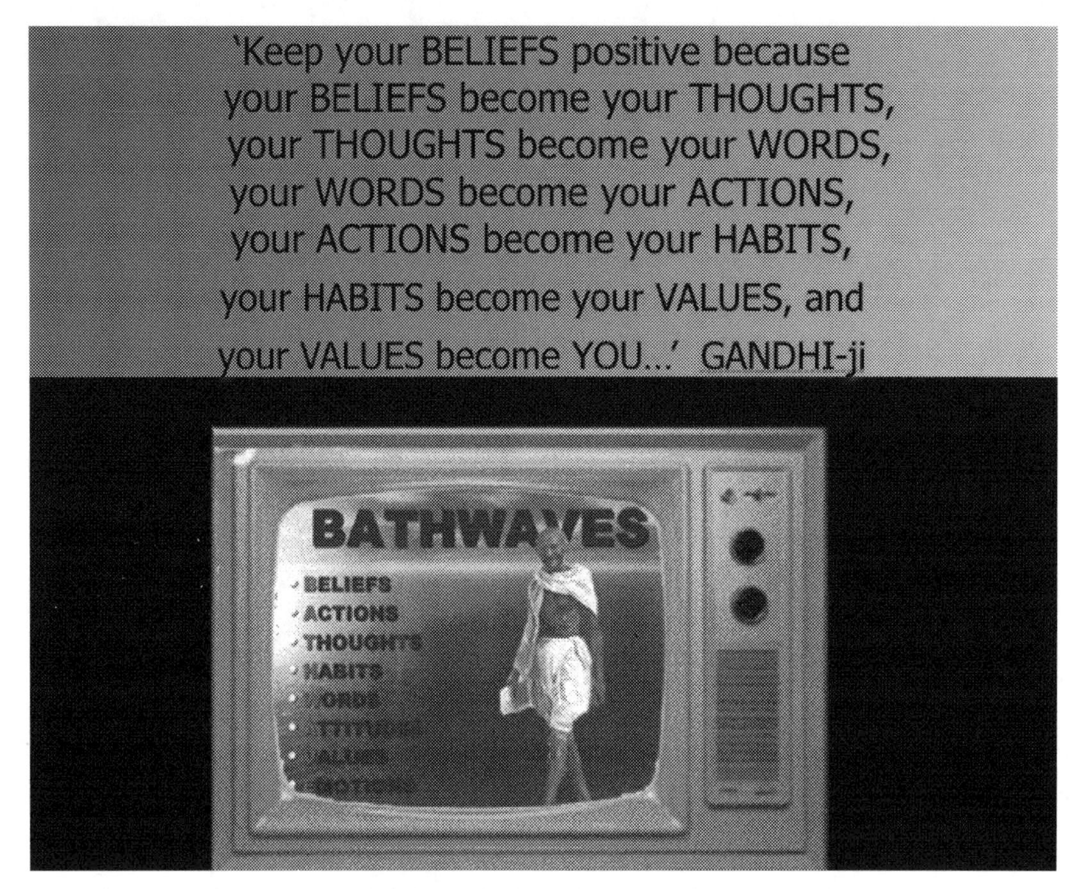

WHAT IS A BATHWAVE?

Besides being the acronym for the informotion of: **beliefs, actions, thoughts, habits, words, attitudes, values, emotions,** it was also the big clue to the technique that became my bridge to quantum wave living.

How do I feel when I take a bath, shower, or swim? I feel good …yet for a limited time. What if I use this BATHWAVE action along with Roger B. Cotting's cup story, as a bridge to speed up my readiness for QUANTUM WAVE LIVING ...because I wasn't able to quantum leap there.

Would I like a quick bathing method to something new?
Tonight we review the technique called the BATHWAVE!

EVENING PRACTICE: THE BATHWAVE TECHNIQUE ACROSS THE BRIDGE TO QUANTASIA

We are now ready for the yellow brick road across the bridge to QUANTASIA.

MORE FUN THAN NUMB

The eight manifestations of my personalbeing BATHWAVEs, can be transformed to LIVING MY DESIRES by using the BENEFICIAL BATHWAVE technique.*

First, I identify BATHWAVEs presenting in the moment, that are obstacles for my well-being. I have listed the first 8 below that drain my energy every time I do them, sometimes, since childhood.

Second, I use BATHWAVE Loving Transformation, (BLT), part 1: facing and embracing my whole body lightly and quickly saying my particular bathwave out loud, I embrace my body lightly, simulating bathing. (see figure 1.6 or watch YouTube videos "Quantum Wave Living" 1-6).

Bathwaving
Part 1: Face and Embrace

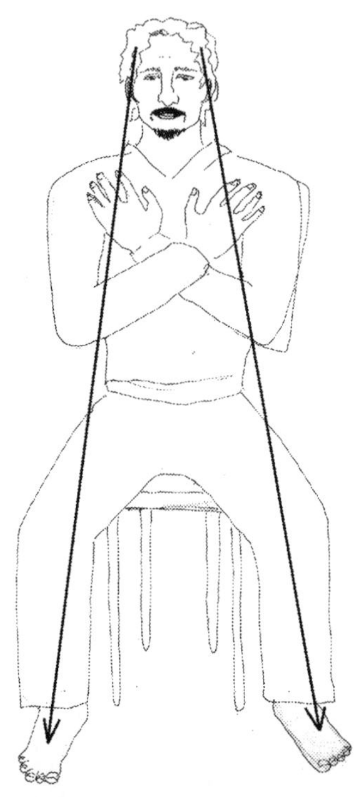

Lightly and quickly hugging the body, saying out loud what you are facing and embracing.

Third, I replace with grace my current BATHWAVE with my desired BATHWAVE, by running my hands around my body's midline, stating "I can, I will, I am (fill in the quantum shift of my desires)". This is the quantum shift stroking motion.

Bathwaving
Part 2: Replace with Your Grace

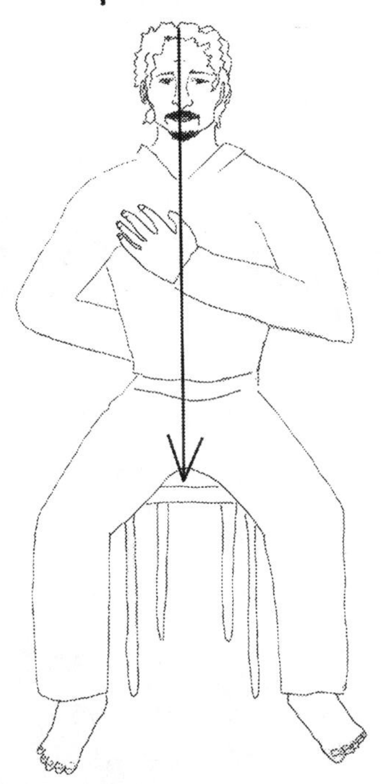

Hands stroke your midline Starting behind your heart.
One hand goes over head to front, down to chair.
At same time, say what you desire in each tense:
I can relax. I will relax. I relax -or-
I can, I will and I relax.

Fourth, I read my new quantum replacements out loud everyday for at least twenty-one days, up to three months. At night I say them with eyes closed and in the morning with eyes open to remember and completely transfigure my current BATHWAVEs to my desired BATHWAVEs. Twenty-one days has been shown to be the minimum number to form a habit and it gently begins my day with a reminder of my emergent new self. This will go on as you continue adding new things to the workbook list at the end of each chapter. Some people like to do the replacements for 3 months! Longer works well when I am able.

Fifth, during my day, I notice whether I am aware of my desired transformations. If not, I may want to redo the technique quickly. BATHWAVEing only takes about 20 seconds as you practice it more.

Remember, some of these patterns have been with me for many years, and it may be easy to slip back into these old patterns if I don't keep awareness and reminding myself.

Am I ready to stop wasting energy with old survival behaviour patterns?
I begin the active process tomorrow in preparation for my quantum wave reading!

QUANTUM TIP:
I AM PHYSICAL QUANTUM WAVES.
I AM MY PATTERNS OF INFORMOTION.
I CAN REPLACE MY INFORMOTION WITH MY TRUE DESIRES INSTANTLY,
MY QUANTUM WAVES CHANGE, SO DOES MY ENERGY AND BODY INSTANTLY IN ENTANGLEMENT.

DAY 3
INTRODUCTION TO
EIGHT BEHAVIOR LOVE LETTERS I GIVE MYSELF, AWAKENING MY RESPLENDENCY

I call these basic eight transformations, the awakening behavior mirror-calls of resplendency.

The first three are related and called the ABCs, which are Accusing, Blaming, and Complaining. These may occupy as much as 35% of my 'resistance' energy that drains myself.
The next three 'lie, hide and deny', can comprise another 35% of my resistance to transformation.

The last two obstacles are defending and justifying making up the rest of my resistance. You will benefit from doing these replacements. I guarantee it.

Let's replace all eight basic awakening behavior mirror-calls (miracles) quantum bridge style... beginning today...if I remember any dreams tonight please record them to work with later. Dreams are the golden communications of self illumination.

Bridge of the Eight Bottom Line Mirror Calls of Resplendence.

ACCUSING/JUDGING

MORNING PRACTICE:
Accusing and/or judging is a pattern rampant in our languaging that will keep us unawaringly accusing and/or judging most of our lives.

It took me years to understand and shift out of this function. Finally, I learned about Parmenides, a Greek philosopher, just before the time of Plato who, paraphrasing, was far ahead of himself when he said:
Humankind creates words that don't even exist, and then spend their whole lives struggling against those concepts that don't exist.

He filled a book with a beginning list of such words. For instance, "bad" was one example of a word that did not exist according to Parmenides. When my mentor, Roger B. Cotting, asked me to find a thing called "bad" that I could hold in my hands, I couldn't. Parmenides called it a "not-thing", which is "nothing". So 'bad' is nothing. Does not exist.

When I understand 'mirror-calls', they replace the antiquated idea of 'bad', as they support our illuminatedself, which is as it was meant to be. 'Bad and good' are only mirror-calls, which I know are shift-able in me.

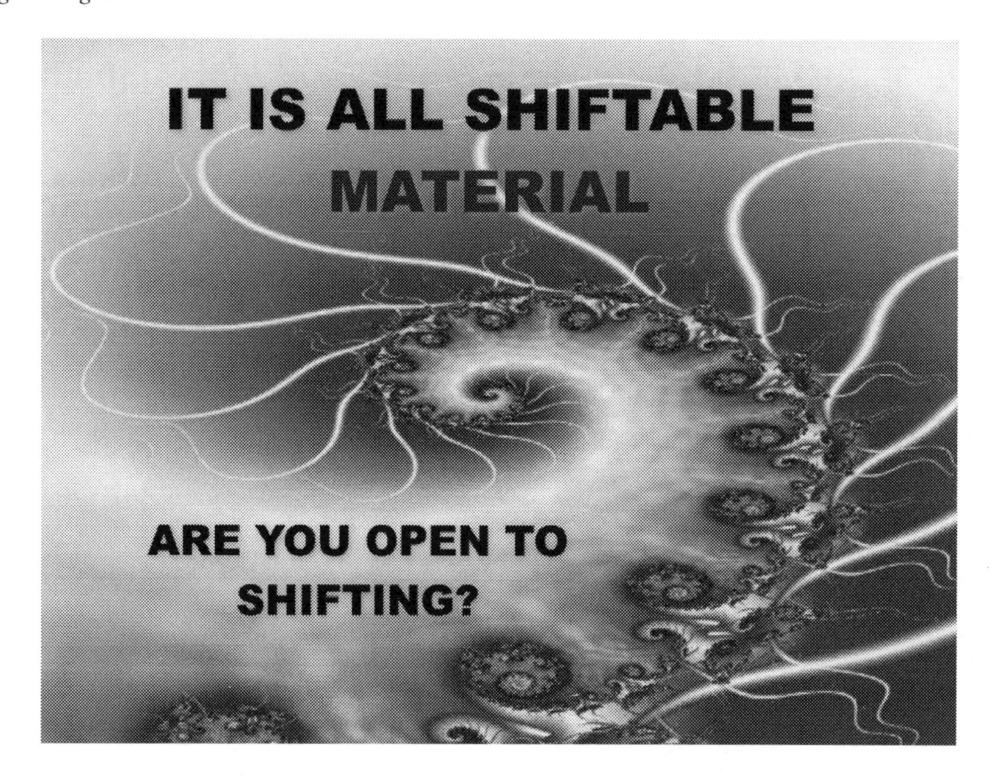

BATHWAVE TECHNIQUE: (This motion is done while simultaneously saying the statement of the BATHWAVE #1 below out loud or whispering in order to access all five senses.)

PART 1:
My hands begin to lightly and quickly embrace (HUG) my body moving from my upper back and my neck over my head and face, down the front of the neck, chest, abs, hips and legs to toes and up the back of legs and side of body to the armpits. Crossing the arms stroking the sides of neck, down the shoulders and arms all the way to the fingertips. And then the back of other hand. Relax the arms, moving the insides across the body as they stroke the body.

When I am in bed I can use the sheets to touch and embrace the body. Use the hands to lightly stroke those parts that are not touched by the sheets such as the head, face, and neck.

PART 2:
One hand paints a midline from my back over my head and face, neck, chest down to the seat, when sitting. The other hand starts on my back, upper spine as far as I can go and moves down to my seat, followed by a slight wiggle, informing the back and bottom front lines to connect creating a full circular orbit.

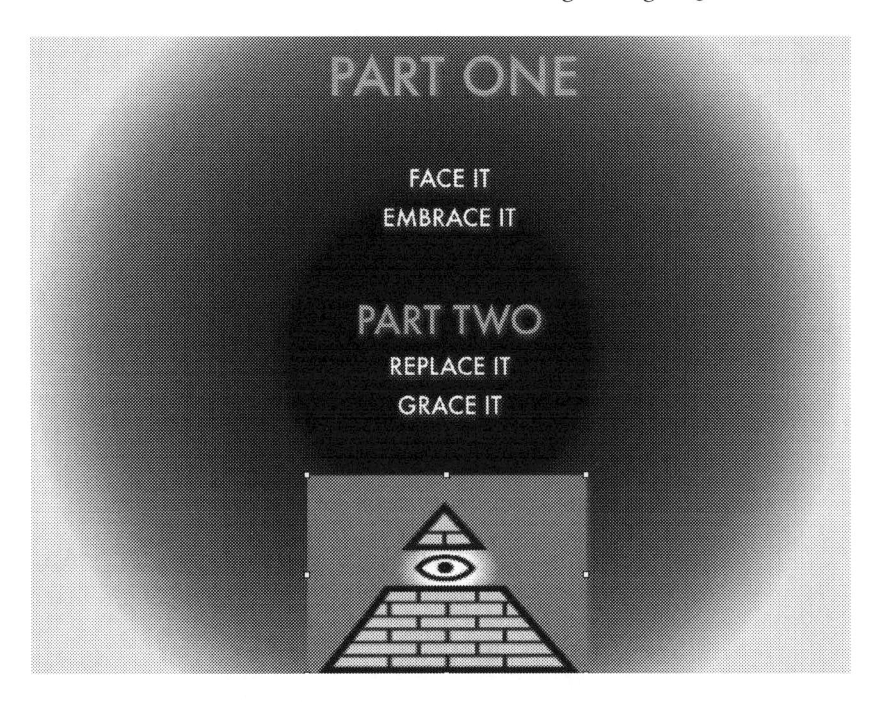

Let's get started!

Part 1: Face and Embrace

Lightly and quickly touch by embracing the whole body as you say out loud, "I face and embrace… (whatever it is) ...and I am ready to replace it."

Part 2: **Replacing with my Grace**

Stroking just my midline all the way around my body and say out loud, "I can, I will, and I am… (whatever I desire to replace it with)"

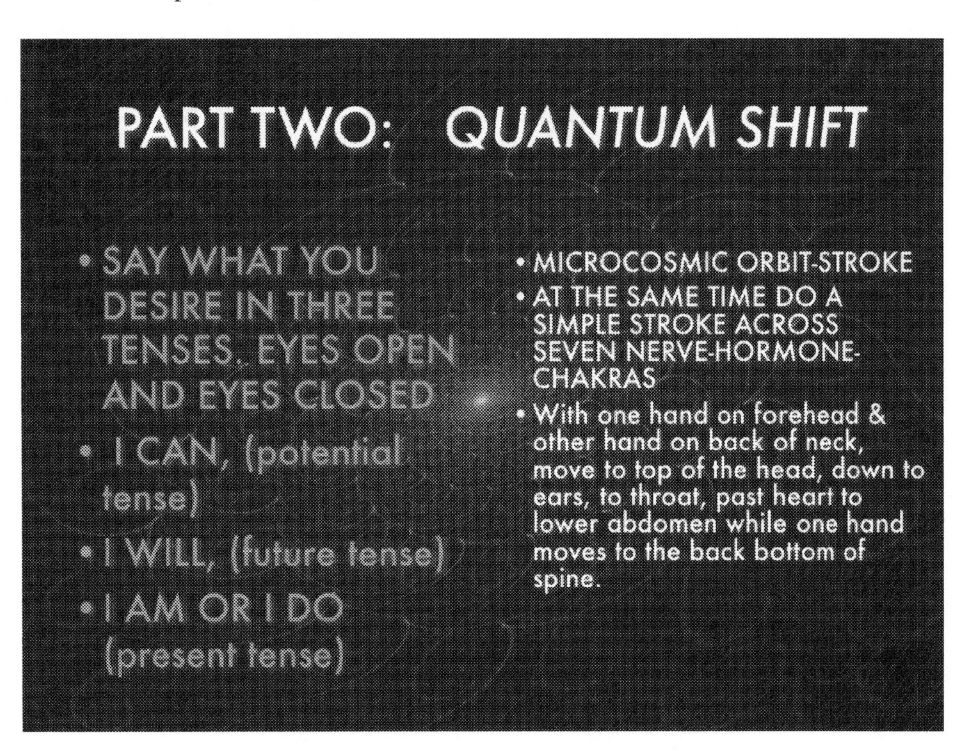

Using can, will, and am helps the brain make this shift. The brain has been pre programmed to think in past, present and future. :-)

Quantum Tips About My Body:

My **back** reveals my "**past**" informotion.
My **front** reveals my " **future**" informotion.
Mv **sides** reveal mv **present**

Do these two BATHWAVES:

Preparing for Nonjudging	
BATHWAVE **Part 1: Face & Embrace hugging whole body**	**Quantum Replacements** **Part 2: Replace with Grace stroking around midline**
I face and embrace that I believe that 'good and bad' exists and I am ready to replace this.	I can, I will, I do believe that good and bad do not exist, only my mirror-calls for my transformation.
I face and embrace that I judge right and wrong.	I can, I will and I am living without judging. I begin to transfigure myself freely.

After each replacement, take a moment, breath in deeply and slowly exhale. Observe my body. Do I notice any changes in emotions, muscles, tension, etc.?

EVENING PRACTICE:
It helps if I see and understand the mirror-call (miracle) nature of my worldly-universe. For as I begin to understand the quantum replacement that 'the universe is benevolent in not interfering with my transformation of these mirror-calls', I begin to appreciate myself in a deeper way.

"Appreciating" is a quantum shift for whomever or whatever I am accusing.
Before I leave this arena, I need to ask whether I am accusing. Please embrace all of these, as it benefits the universe along with me.

Accusing	
BATHWAVEs **Part 1: Face & Embrace**	**Quantum Replacements** **Part 2: Replace with Grace**
I face and embrace I accuse people or things.	I can, I will and I do appreciate people or things.

I face and embrace I believe something is against me or blocking me.	The Universe can, will, and is benevolent and supporting my authenticity.
I face and embrace I feel accused by others.	I can, I will and I am valuable and gentle with myself.

Judging Self and Others	
BATHWAVEs **Part 1: Face & Embrace**	**Quantum Shifts** **Part 2: Replace with Grace**
I face and embrace I judge myself and/or others.	I can, I will and I do live without judging myself and/or others.
I face and embrace I feel judged by myself and/ or others.	I can, I will and I am free to be me. I can, I will, and I love myself unconditionally.

When I am accusing/judging myself it is mirrored by others. When I know the O-ring muscle test I can use it to check my quantum shifts. See appendix if interested. Otherwise, I trust my heartfelt sense of myself. No need to have all the answers, as my sincere desire to transfigure myself will do all the work.

Shifting our BATHWAVEs is the treasure hunt that leads one from accusation to self-actualization.

*(For muscle testing and 'noticing': The quantum shift that I need to put in will feel needed or muscle test weak.. After the quantum shift is put in, it tests strong. If it doesn't test strong, I have something else blocking it, which I can shift. Then I can go back to test the original.)

DAY 4
BLAMING

MORNING READING:
Blaming is an obvious one, and it comes with the territory of modern media, which only reports the struggles and conflicts in the world. When I embrace that there's no need to blame, just to understand the nature of reality, blaming can be eased and I can replace it with 'blessing'.

When I Face and Embrace "I am blaming others for the way things are," I might replace it in Part 2, with the quantum shift, "I bless others for being a mirror-call of my resplendent nature."

When I feel blamed which would be the mirror of self-blame, the quantum shift might be: "I understand and bless myself." Let's replace these now or this evening if easier.

Blaming	
BATHWAVEs **Part 1: Face & Embrace**	**Quantum Shifts** **Part 2: Replace with Grace**
I face and embrace I am blaming others.	I can, I will, and I do bless others for the way they are mirror-calls for my resplendency.
I face and embrace I am blaming myself.	I can, I will, and I do understand and bless myself.
I face and embrace I feel blamed.	I can, I will, and I do feel blessed.

After each replacement, take a moment, breath in deeply and slowly exhale. Observe my body. Do I notice any changes in emotions, muscles, tension, etc.?

EVENING PRACTICE:
Say my quantum replacements in present tense only, out loud before sleep. Read first then say with my eyes closed.

DAY 5
COMPLAINING, CRITICIZING, AND COMPARING

MORNING READING:
This one is occupying the largest percentage of my life, and can be embraced again and loved, as a less than empowering habit of my daily routine. Instead, I can 'connect' and 'communicate' with myself and others until I empathize with the person I was complaining about. This person is sometimes 'me' or a mirror-call (miracle) of me.

My quantum shift might be that I desire to understand what a person is feeling and their unmet need. Relationshifting these awakening mirror-calls profoundly transforms relationships without the need of confrontation.

Complaining, Criticizing and Comparing	
BATHWAVEs **Part 1: Face & Embrace**	**Quantum Replacements** **Part 2: Replace with Grace**
I face and embrace that I am comparing, complaining, criticizing myself and/or others.	I can, I will, and I am listening, connecting, and communicating. (empathizing)

EVENING PRACTICE:
Say my quantum replacements twice out loud before sleep, once with my eyes open and once closed.

DAY 6
LYING

MORNING READING: Say my quantum replacements in present tense only, out loud with my eyes open.

These next three bottom lines are easy to remember because they have a little rhyme within them. Lying, Hiding, Denying. Lying is really about deceiving my self (and/or others) about the way I believe things really are. The quantum shift is that "I am open and honest with myself at all times" which serves me to a much greater degree.

Lying	
BATHWAVEs **Part 1: Face & Embrace**	**Quantum Shifts** **Part 2: Replace with Grace**
I face and embrace I deceive myself.	I can, I will, and I am honest and open with myself at all times.
I face and embrace I deceive others.	I can, I will, and I am open and honest with others with discernment.

EVENING PRACTICE:
Say my quantum replacements out loud before sleep, with my eyes closed.

DAY 7
HIDING

MORNING READING: Say my quantum replacements in present tense only, out loud with my eyes open.

Hiding from my self and others being invisible, and not being seen or heard, can be a tactic of survival for children and adults. As I enter adulthood I can embrace this, realizing that I no longer need to hide my voice or my being. I can speak my desires and step up to the plate however I choose, and those would be worthy quantum shifts. "I am visible and speak my desires."

Hiding	
BATHWAVEs **Part 1: Face & Embrace**	**Quantum Shifts** **Part 2: Replace with Grace**
I face and embrace I am hiding from myself/ others.	I can, I will, and I am visible and speak my desires.

The Relationshifting method would have understand that any hiding I do with others, I probably do with myself also.

EVENING PRACTICE:
Read and say my quantum replacements out loud before sleep, with my eyes closed in present tense only.

DAY 8
DENYING

MORNING READING: Say my quantum replacements in present tense only, out loud with my eyes open.

This movement of denial has created the 'unconscious' of Western psychology. Since I don't really want to know everything, I 'unconsciously' deny things from my awareness without knowing it.

With courage, honesty, and commitment, the unconscious does not really need to exist. I can embrace this denial function, replacing it with the quantum shift that "I am aware of everything about me."

The unconscious action of others is not my business unless I react or judge them. This means they are mirroring me. Then I need to be clear to what I am reacting. Then use Relationshifting to see what needs to be shifted, and Bathwave what it says about me. This will be explained in detail in Decade 4.

This may or may not feel good at first, yet it does as soon as I face and embrace things, which takes only seconds. Bathwaving my eight bottom lines is in the long run desirable for my well-being, my wisdom, and my relationships.

I have heard it said that what is in my unconscious is what usually kills us. I had at least 15 forms of anger in my unconscious with my liver cancering. Whew! And I still Bathwave the triggers behind those emotions sometimes.

For those who know how to muscle test myself, I cannot muscle test these eight Bottom line Behaviors in the normal way for myself, because I might lie, hide or deny about it and give myself the wrong answer!

There is a method I devised using the circular fingers muscle testing, to determine them, say, whether I am in denial. I choose a name of someone whom I trust, and ask if I can 'ask about me from their position' for a moment, muscle testing for the answer about me. This can get the correct answer for me.

For instance, when I muscle test I say, "May I test as Jessica?" I get a yes. Then I muscle test the question as if I were Jessica, "Is Angela in denial?". I get an accurate answer. If "yes", I need to 'embrace' this denial out loud with the final quantum shift, "I am aware of everything about me". See below.

Denying	
BATHWAVEs **Part 1: Face & Embrace**	**Quantum Replacements** **Part 2: Replace with Grace**
I face and embrace I have been unknowingly burying things under an "unconscious" rug and I am ready to replace this.	I can, I will and I am aware of everything about me.

EVENING PRACTICE:
Read and say my quantum replacements out loud before sleep, with my eyes closed.

DAY 9
DEFENDING

MORNING READING: Say my quantum replacements in present tense only, out loud with my eyes open.

When I feel or believe that I am being attacked, I will go into defense mode. Lawyers and judges are very, very good at this. They are paid to be. I do not ever need to defend myself, unless I am in that field or career.

I can embrace that I tend to defend that which I believe in, but I realize it's a waste of my energy. I might put in, "It's fine to be just the way I am" or "I enjoy being the way I am" or "Everything is good, just as it is."

Defending	
BATHWAVEs **Part 1: Face & Embrace**	**Quantum Replacements** **Part 2: Replace with Grace**
I face and embrace I am defending myself and/ or others	I can, I will and I am fine just the way I am.
I face and embrace I am explaining myself and/or others	I can, I will and I do enjoy being the way I am without explaining.

EVENING PRACTICE:
Say my quantum replacements out loud before sleep, once with my eyes closed.

DAY 10
JUSTIFYING

MORNING READING: Say my quantum replacements in present tense only, out loud with my eyes open.

This is our response to believing I am being judged or attacked. Nobody needs justifying. Some people get paid for justifying: managers, CEO's, people who run things are paid to justify and do cost benefit analysis. However, in my life I do not need to do that, outside of the field of business.

In my daily life, energy can be depleted justifying what I am doing, telling others, or myself about my justification. Let's embrace justification and put in "I do as I desire," or "It's ok to live my desires."

Justifying	
BATHWAVEs **Part 1: Face & Embrace**	**Quantum Shifts** **Part 2: Replace with Grace**
I face and embrace I am justifying and proving myself to others.	I can, I will and I do as I desire. I am living my true desires.

There is no proving in the quantum world as Heizenberg noticed; my ability to 'prove' is uncertain. There is no need to prove yourself or anything...as they say, "the proof is in the pudding".

Notice my day without justifying or proving.

EVENING PRACTICE: Notice any difference without justifying or proving.? Write about it in my notebook or with my computer.

These eight obstructions or diminishings of my energy or chi, once replaced, open the way for a deep practice of Relationshifting. This is my invitation to the rest of my quantum living... health, wealth, wisdom of my relationships supporting my illuminatedbeing.

The most important is my basic relationship with my triune self: personalbeing and quantum wave being communicating, from which emergent is my resplendent illuminatedbeing.

If I have difficulty shifting any or all of them I am not alone. I give myself twenty-one days to three months of repeating my eight desired bottomlines. With my active awareness they will diminish and I will be able, for example, to REPLACE justifying to "doing what I desire" in a moment.

The eight manifestations of my personalbeing, BATHWAVEs, are transformed to MY DESIRES by using the beneficial BATHWAVE technique.

CONGRATULATIONS, I COMPLETED DECADE ONE!

SUMMARY:

First, I identify these Eight Bottomline behavior BATHWAVEs presenting in the moment that are mirror-calls for my well-being.

Second, I use BATHWAVE Loving Transformation, (BLT), to face and embrace my current BATHWAVEs using all five senses. See the BATHWAVE technique described above from Day 3.

Third, I replace with grace my current BATHWAVEs with my desired BATHWAVEs by stating "I can, I will, I am (filling in the quantum Shift of my desires) running my hands around my body systems midline.

Fourth, I read my new quantum replacements everyday for twenty-one days, at night with eyes closed and in the morning with eyes open. This action completely transfigures my current BATHWAVEs to my desired BATHWAVEs.

Fifth, every few days, I can notice whether I am holding the transfiguration or I can use any muscle testing technique or pendulum etc. If notholding then I quickly replace that BATHWAVE again.

I remember, that some of these patterns have been with me for many years, and it may be easy to slip back into these old patterns. Please be gentle and just repeat them.
I still repeat denial and judging regularly!

As I become more aware of my resplendent BATHWAVEs, I see basic awakening mirror-calls, that create resistance to my ability to live my true BATHWAVEs. These eight mirror-calls are important, as they help me see what I desire to transform.

Congratulations! I am ready to go with my quantum glow. I am discovering my quantum wave pattern in this next Decade which is my encode-decode key living my full awareness as my unique **QUANTUMCODE being.** This is a pivotal tool.

QUANTUM REPLACEMENTS FOR MY EIGHT BOTTOM LINES

I repeat these present tense replacements for at least 21 days to 3 months (highly recommended). Before bed say with eyes closed. In the morning say with eyes open.

I believe that good and bad do not exist.

I am living without judging. I begin to transfigure myself freely.

I appreciate people or things.

The Universe is benevolent and supporting my authenticity.

I am valuable and gentle with myself.

I live without judging my self and others.

I am free to be me.

I I love myself unconditionally.

I bless others for the way they are mirror-calls for my resplendency.

I understand and bless myself.

I feel blessed.

I am listening, connecting, and communicating (empathizing).

I am honest and open with myself at all times.

I am open and honest with others with discernment.

I am visible and speak my desires.

I am aware of everything about me.

I am fine just the way I am.

I enjoy being the way I am without explaining.

I do as I desire. I am living my true desires.

Throughout the book I will receive daily friendly reminders to continue reciting my quantumwavecode (from decade 2 later) and replacements. Saying my quantumwavecode and replacements are a vital aspect of Quantum Wave Living, which is why I am receiving encouragement each day to do so.

DECADE 2

READING AND NAMING MY QUANTUM WAVE-LIKE PATTERN CALLED MY QUANTUMCODE

QUANTUMCODE HAIKU

What am I wishing
To do? My quantum wave code
Shows my glowing clue!

My quantum wave-like pattern is my unique, self-chosen, group of pulsating actions of informotion that I am living ...and that I might be resisting.

This quantumcode of me is my secret to GLOW. Have you heard the old saying, "go with the flow"? The quantum wave living rendition of it is "go with my glow"! The big difference is that it goes in ALL directions and instantly communicates (by entanglement) throughout the world and my living.

In this Decade I will read my **unique abidingself, which is my QUANTUMCODE key** for awarefully living my true desires!

My quantumcode self =
my quantum wave-like pattern =
my abidingself (meaning: committing to my personalself) = accompanying
my eternalself (meaning: my recurring pattern throughout my living and throughout the universe, recorded in quantum memory forever).
These words are used interchangeably throughout this book to help me remember the qualities of my forever companion.

My EB, Eternalbeing is my unique quantum wave pattern recurring throughoutmy quantum waves pulsating as vast as the universe.

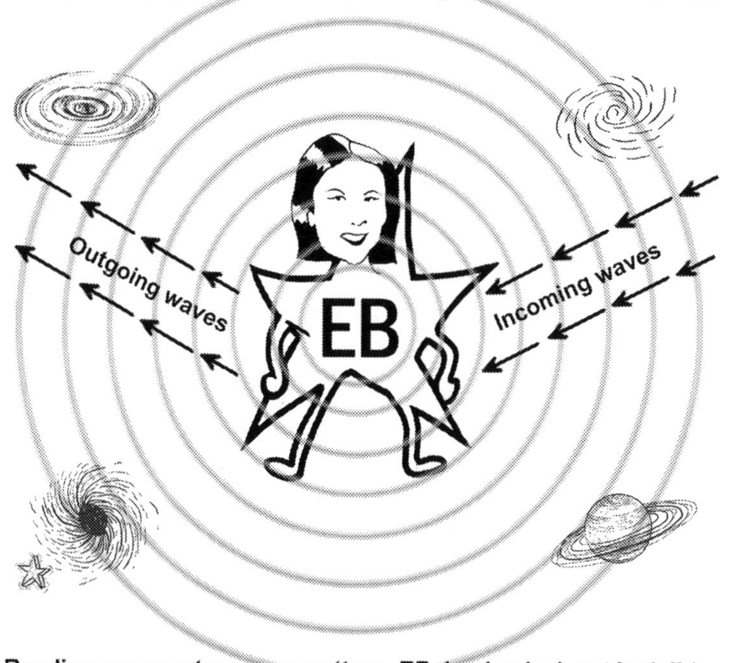

Reading my quantum wave pattern, EB, is physical, yet invisible to my eyes, communicating and giving me access to all information in googleverse.

DAY 1
MAKING A LIST OF 6 THINGS I DID THAT I WAS GLAD AND/OR PROUD OF WHEN I DID IT

MORNING PRACTICE: Make a list of six specific things I have done in my whole life, which I was proud of. Not proud of myself, but proud of the thing I did. It has nothing to do with the size of the action, but the size of the gladness when accomplished.

Write those moments down as they come to me, not necessarily in chronological order. If I have difficulty remembering them, do it throughout my day, keep my notebook with me. They need to be things that I was passionate about doing.

Any six specific things I did in my whole life that I felt proud of when I did them (simple or complicated):

1)
2)
3)
4)
5)
6)

EVENING PRACTICE: Complete my list of 6 accomplishments I am proud of.
Say my Decade One quantum replacements out loud before sleep with my eyes closed.

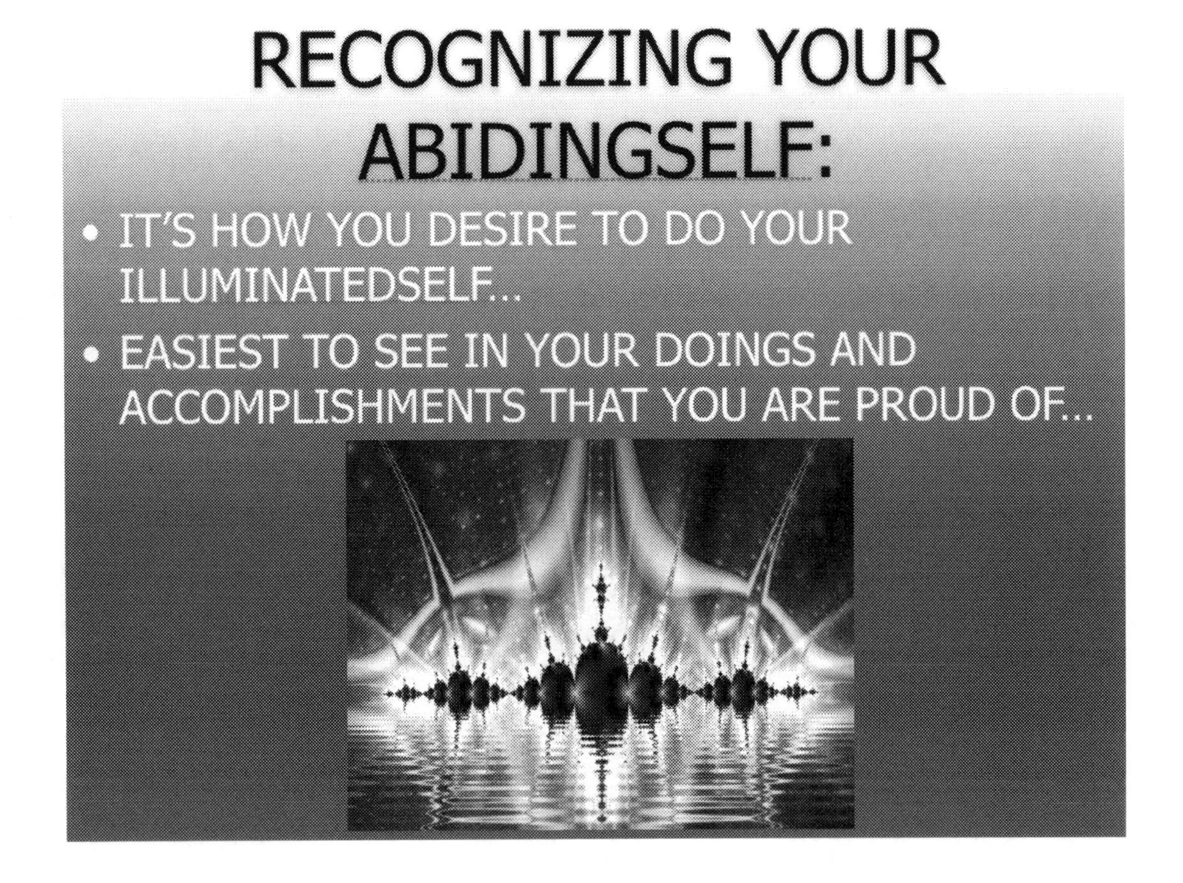

DAY 2
SPECIFICALLY, HOW DID I DO MY FIRST ACCOMPLISHMENT?

MORNING PRACTICE: Say my replacements with my eyes open.
I need 20 to 30 minutes to do this process, because it is important.

Part 1: How did I accomplish this? ***What actions were required to accomplish it?***

Find new words or expressions that describe the big picture of what I did and how I did it. Say nice things about myself as if it's for a newspaper article. Add these words to the list I began in Part 1.

Part 2: Why was it important to me?

Pretend I am a journalist interviewing myself about the first thing I did on my list of accomplishments written yesterday morning.

I ask myself, "**Why was it important that I did this?**" and write down in list format, any things that come up in concise language. They don't have to make sense. Stop reading this and begin writing. When I have written all that I can proceed to part 2 below.

I am looking for action words, **verbs**!

Keep writing until I have a minimum of five verbs (action words) or expressions.

Keep this in mind during my day and write down any new words which come to me.

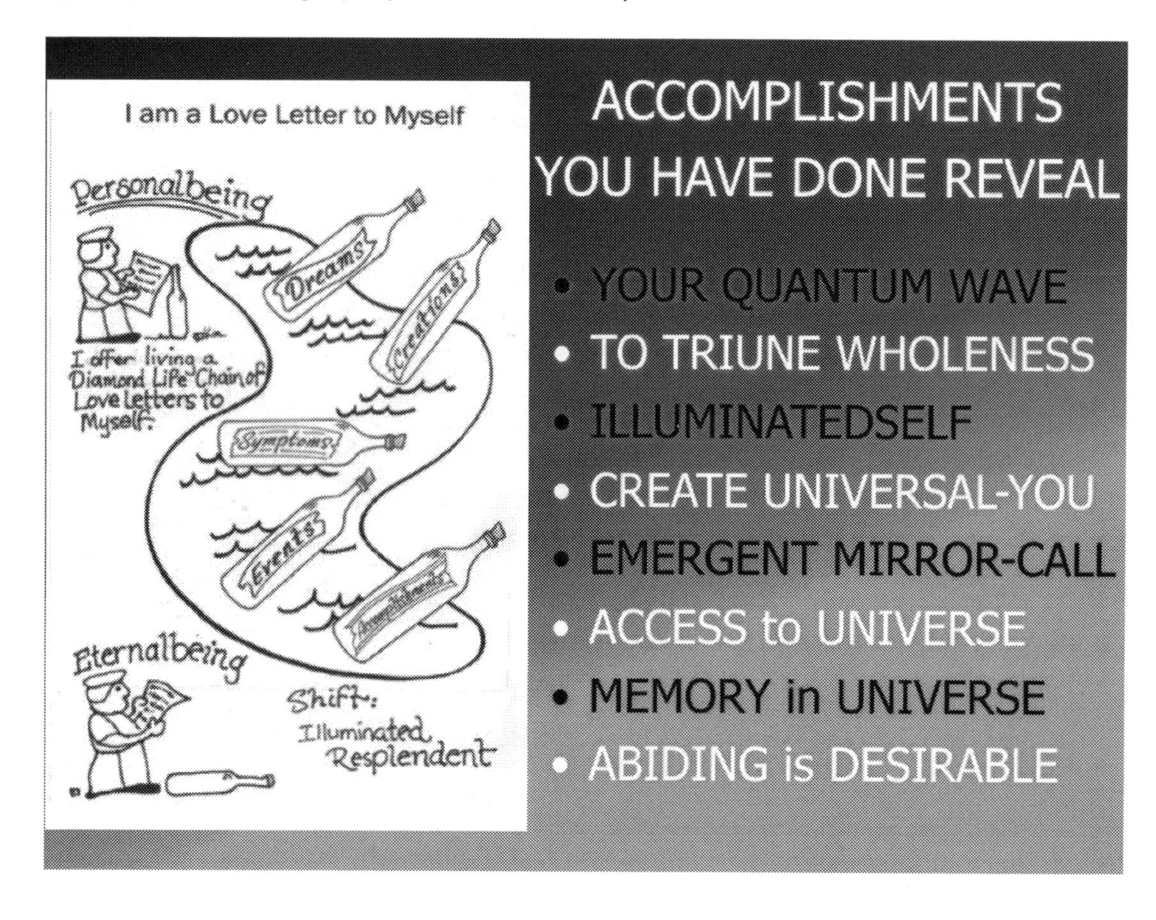

EVENING PRACTICE: REWRITING MY LIST

Here are the guidelines for rewriting my list:

1. Rewrite all the verbs in present tense. Example change "visualized' to "visualize". If I have trouble doing this, insert the word 'today' early in your sentence read out loud and the correct verb will come. Write it down.
2. For any people in your list, you need to substitute what aspect of you they represent: see relationship table in Decade 4 or appendix.
3. Take out judging and feeling words like bad, wrong, good, right, etc.
4. For any specific objects listed look for equivalents in the quantum descriptionary in Decade 5 table_ in appendix. Example: for "house" replace with "self", for 'car' replace with 'physical body'... use the word 'thing' or 'tool' when it is not on the list.

It's ok if one or two items on my list are not actions. Sometimes the words may picture an action without being an action. If I have more than five actions, circle the five that are alive or important to me. Or ask my best friend or family member who is supportive and knows me well. Next, randomly number these words or expressions from 1 to 5, and write them in the sentence framework below.

My quantumcode self (quantum wave-like pattern) is

1) ____

2) ____

3) ____

4) ____

5) ______________________________ as my whole radiant illuminatedbeing.

Read this sentence three times out loud, rewrite any changes that come during this process of reading to help it sound like me. Don't worry if it doesn't make sense as you have as long as you like to flesh this out.

Don't worry if I am unsure of the order of the words in the sentence, because I will repeat this process with a new random order of the chosen five words and expressions for the next four days. It gets easier as I begin making more sense of the expressions and glow flow of my quantumcode.

Before sleep: Say my Decade One quantum replacements once out loud before sleep with my eyes closed. Read the quantumcodebeing sentence I wrote this evening.

DAY 3
REARRANGING ORDER OF QUANTUMCODE

MORNING PRACTICE: Say my Decade One quantum replacements once out loud with my eyes open. Read the quantumcodebeing sentence I wrote last evening.

EVENING PRACTICE:
Randomly assign new numbers for my chosen five words and expressions and repeat the process of writing and reading it out loud until you make new sense of it.

Before sleep: Say my Decade One quantum replacements once out loud with my eyes closed. Include reading the NEW sentence I wrote this evening of my abidingself.

Day 4
ADD NEW WORDS ONLY WHEN YOU FEEL GUIDED

MORNING PRACTICE: Read out loud my quantum replacements and my abidingself statement from yesterday with eyes open.

Once again, assign numbers for my chosen five words and expressions in a different order and repeat the process of writing and reading my quantumcode out loud to make sense of it. Note any edits, only add new words if I am strongly self-guided to do so.

EVENING PRACTICE: Say my Decade One quantum replacements out loud before sleep with my eyes closed. Read the new sentence I wrote this morning of my abidingself.

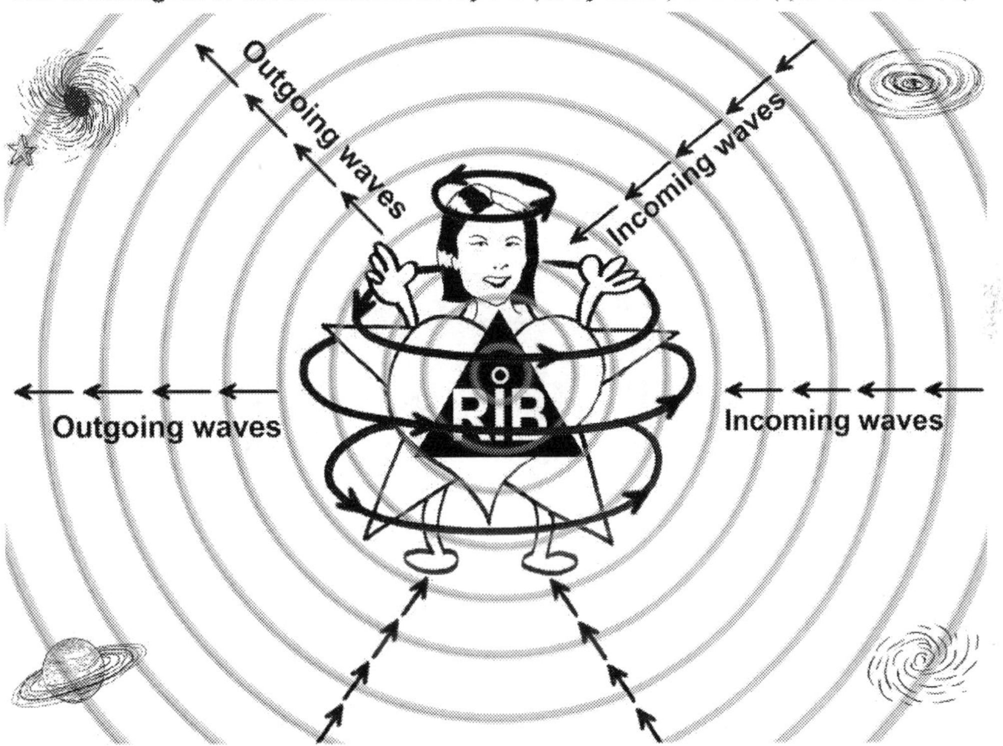

DAY 5
INSTANT QUANTUM LEAP HIGH-QUEUE

Living homemade
Quantum wave being, pops out
Emergent miracles!

MORNING PRACTICE: Repeat the process from the prior morning of reading my quantum replacements. Begin to choose numbers for my chosen five words and expressions and repeat the process of reading out loud to make a new sentence, and write it down.

EVENING PRACTICE: Say my Decade One quantum replacements out loud before sleep with my eyes closed. Read the new sentence I wrote this morning of my abidingself.

DAY 6
DESCRIBING MY ABIDINGSELF

MORNING PRACTICE: Great job… repeat the process from the prior morning of reading my quantum replacements.
Today is the last day to write a new sentence describing my abidingself with the same words in a new order.

This will be another working eternalself for the next fun processes.

EVENING PRACTICE: Say my Decade One quantum replacements out loud before sleep with my eyes closed. Read the sentence I wrote this morning of my abidingself.

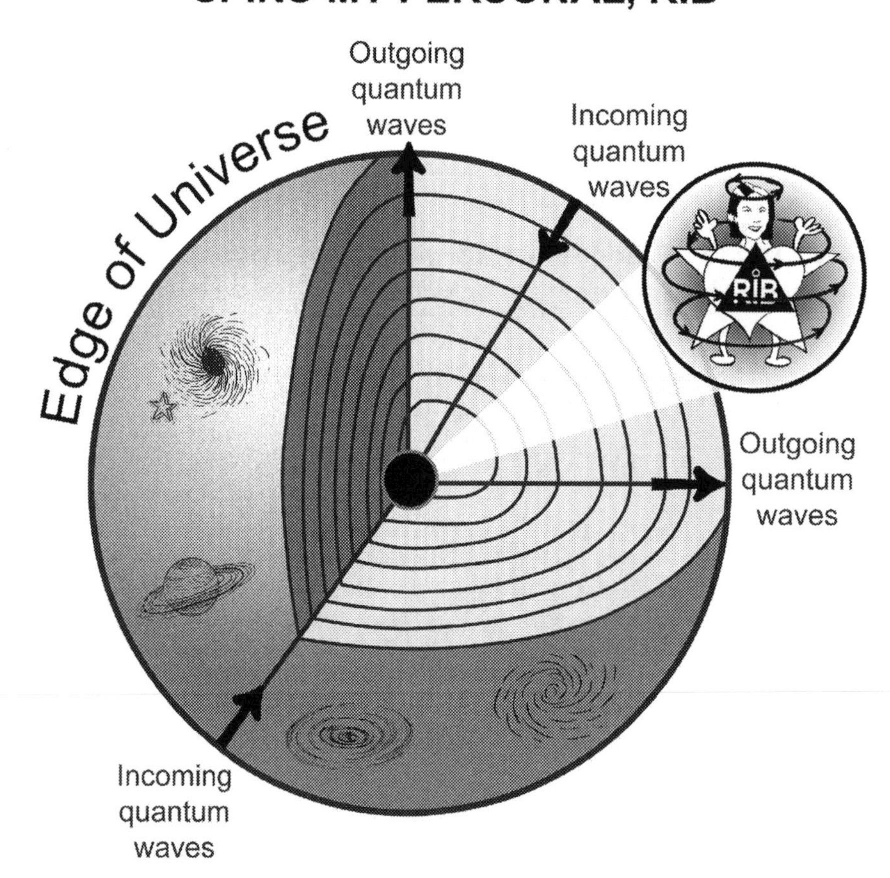

DAY 7
CHECKING YOUR ABIDINGSELF CHOICE

MORNING PRACTICE: PICKLING MY ABIDINGSELF

After reading my quantum replacements out loud with my eyes open, I am ready to choose the crispiest of my five quantumcode pickles.

Choose the sentence that best summarizes my experience of my first accomplishment.

Ask myself the question, "would I be happy doing this for the rest of my life?" Read each of the five sentences I wrote, until I can clearly say yes to one of them. This is a challenging part of this whole quantum program, yet please persist because it's worth it!

Understand there is no right or wrong choice. It's important to pick the one that resonates with me. This choice may change later.

If I have trouble choosing one of the sentences, record reading them out loud all in succession, numbering each one. Listen to them repeatedly until one of them calls out to me.

If I absolutely can't decide, then use a coin toss. There are no accidents just keep tossing the coin. Read down the list, heads is yes and tails is no. Continue doing the process until I get down to one.

EVENING PRACTICE: CHECKING YOUR ABIDINGSELF CHOICE

The Next Fun Process I Will Begin:

Read the second accomplishment on my list. Next, read my abidingself sentence, asking myself if these actions were present in any way shape or form in accomplishing this one also? Be creative in understanding myself this way. I am sure the answer will be yes.

Do the same process with all four remaining accomplishments on my list. If I do not get a "yes" for all of them, question whether it was really me that was proud of the accomplishment or was it someone else's pride that I identified with?

Eventually, **I may realize that this quantumcode key is secretly a large part of the proudest moments of my living.**

It is actually my pattern of the quantum wave-like me, which is as large as the universe. And it is important to the universe for its manifestations and therefore its expansion!

I am seeing how I have been doing this quantumcode pattern whenever I felt proud throughout my life. **The key now is to learn my abidingself pattern and begin practicing living this way, with awareness, in everything I do**. Then I am beginning to live my true desires in each moment. You will only know you are doing it in hindsight.

Adding the abidingself to my quantum replacements list to repeat every night and morning inspires my day!

Read it with my Decade One quantum replacements and then repeat out loud with my eyes closed.

Looking through Heart Shaped Glasses
is my Personalbeing Particularizing.

Looking through Star Shaped Glasses
is my Eternalbeing Visioning the Recurring Me.

Looking through Heart/Star Shaped Glasses
is my Resplendent Illuminatedbeing Living
the EEEZY Quantum Resplendency.

DAY 8
NAMING MY ABIDINGSELF WHICH WILL BE MY NEW SECOND NAME

MORNING PRACTICE: NOW GET MY MAGAZINES, CHALKS, CRAYONS OUT TO MAKE A POSTER AND MEMORIZE IT IN A SIMPLE FUN WAY

Say my Decade One quantum replacements once out loud with my eyes open. Read the quantumcode-being sentence.

Choose a name that reminds me of my abidingself and what he or she does; a word from the sentence is fine. Time to get my art supplies out and make a poster of my written abidingself quantumcode.

My personal favorite is to tear or cut pictures from magazines that remind me of aspects of my quantum wave pattern. Make a collage with photos, symbols, colors or tearings that remind me of my quantumcode self. If you can't find poster board or large table paper, use old newspaper or large paper bags. Tape paper and pictures on it, glue dots work well.

I place my new quantumcode name at the top after my personal name. My quantumcode name may change over time until one feels correct.. You can tape the new name over it. Decorate my poster, using colors, paper, paints, crayons... to enhance the visual effect or not. Have fun, as I deserve it after accomplishing this process. I will complete the poster this evening or whenever I get my images. I plan to write the words of my quantumcodebeing on or around the pictures so you can read them everyday.

EVENING PRACTICE: Finish my poster of my abidingself. Put it up in a spot I look at often.

Say my Decade One quantum replacements out loud before sleep with my eyes closed. Say the sentence of my abidingself, by heart when memorized.

DAY 9
COMMUNION IN ONE-NEST!!!

"If I want to maintain my eternalself (abidingself) active in my life, be careful not to integrate or blend my eternalself (abidingself) and world into my life today! ...it is vital to communion the two, ...not merge them ...nor unite them into a oneness! **The personal self and eternalself (abidingself) need to live in one-nest together**...and certainly do not separate them!" Cotting, R. B. (2018) "14a Pericope

MORNING PRACTICE: Say my Decade One quantum replacements and quantumcodebeing sentence once out loud with my eyes open.
My abidingself is a quantum wave-like pattern of my true desires eternally.

I take my written abidingself with me today and read it as often as possible. If I have a demanding day I take it with me anywhere, even the toilet! I want to memorize it by tonight if possible.

EVENING PRACTICE: MAKING A RECORDING

I use my smart phone, computer, or other recording device and possibly in the background, soft music without words. I record myself reading my quantumcode in short segments. I pause between each segment long enough to repeat out loud the segment. I also read the Decade One quantum replacements in a similar manner with a pause after each. I end the recording repeating my abidingself pattern. Take a deep breath in the pauses between reading sentences. This brings real energy to the sound. Recording is a powerful tool to use throughout this journey.

If I don't have a recording yet, I keep reading my Decade One quantum replacements out loud before sleep with my eyes closed. Do the same with my abidingself.

DAY 10
YOUR FOREVER PARTNER

"As I awaken and experience my eternalself (abidingself)...<u>do not surrender my personalself or try to become one with my eternalself!</u> (abidingself)<u>...for I am to function as partners and companions as I decide</u>...or not at all!" Cotting, R. B. (2018)

MORNING PRACTICE: When I awaken say my Decade One quantum replacements and quantumcodebeing sentence once out loud with my eyes open.

How would I like to have another unique individual forever partner with me, who is only there performing my sacred true desires in each moment?

My quantum wave-like pattern I call my eternal abidingself and quantumcode is:

"... my declaration of how you will be eternally alive and is a particular yet vast self-chosen blueprint of my destiny! ...the universe does not care for you as you should be! ...it functions to care for you as you truly desire!" Cotting, R. B. (2016)

This means the universe supports my intimate personal relationship with my true desires, which is living my abidingself. Emergent from this interaction is my universal whole illuminatedbeing.

<u>**Today, "Smile to my abidingself" all day. This means to smile at people as if I am smiling at my abidingself. As I am living it, they are mirroring it.**</u>

Think of my abidingself name when I am alone and smile as if I fell in love for the first time.

Practice saying my abidingself statement in each free moment.

EVENING PRACTICE:
Using the automatic writing technique, (writing continuously without pausing to think or edit), write a love letter of my wishes to my quantumcode self. Tell them my truth today whatever it is. This letter is just for me.

Before sleep, if ever I know ahead of time that I cannot read my quantumcode and replacements for that 5-10 minutes in the morning I can do this:
Say my quantum replacements twice out loud from Decade One, once with my eyes open and once closed. Do the same with the sentence I wrote of my quantumcode eternalself, when possible by heart, with eyes open at the beginning and closed at the end.

I congratulate myself for completing the most important and demanding Decade of the process.

Completion: In Decade 3 I begin awakening my ability to vision and live with my abidingself. If I have time I can read the introduction tonight.

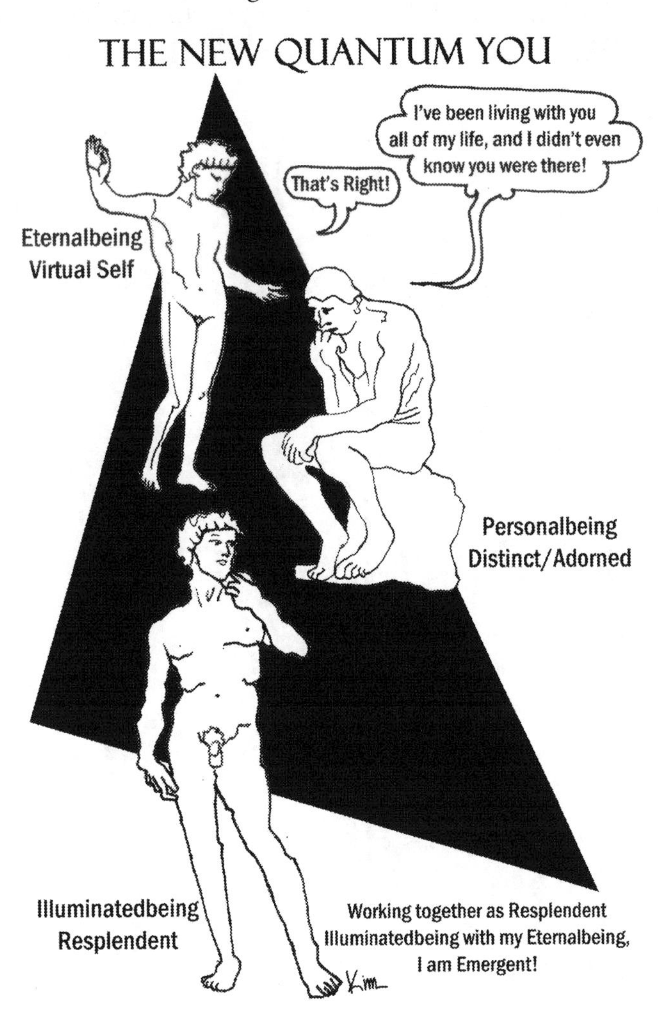

DECADE 3

MAGINATION:
LIVING MY QUANTUM WAVE PATTERN IS MY MAGIC-NATION:

"I AM A QUANTUM MOVIE PROJECTOR, NOT A CAMERA; EVERY SCENE IS SELF BEING SEEN."
Angela Communion-ah

I am a movie projector, not a camera! How do I make my movie?

The story line theme of my movie is my quantum wave pattern called my eternal abidingself. My material that guides the scenes are my beliefs, actions, thoughts, habits, words, attitudes, values, and emotions. (As I take each of their first letters, and I get the acronym BATHWAVE).

My film star 'personalself' is taking a BATHWAVE with my abidingself and it's recorded automatically by Universal studios. Anyone can watch it for free, forever. And it gets even better than that!

All I have to do is add new information that I desire in my life. My movie automatically gives me surprise miracles (mirror-calls*) that are desirable by ME! So my series is always changing in a new way that satisfies me. And this is happening this moment. IS THIS GREAT ENOUGH?

WHEN I don't add new information or WHEN I AM NOT awarefully IN CHARGE OF MYSELF, THEN I AM LIVING RERUNS.

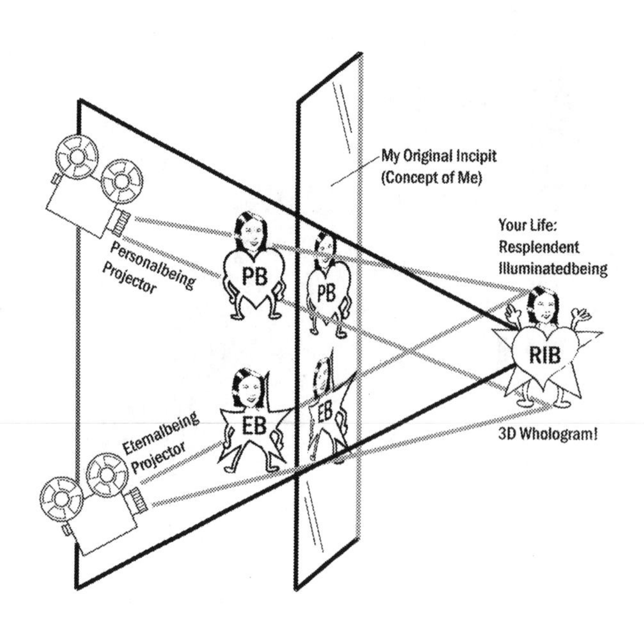

AM I MAKING RERUNS?

Reruns can get very tiring especially after watching them for many Decades over and over and over.

Don't worry because now I know my abidingself eternally from Decade 2, which are the actions I do throughout my life, a recurring pattern, like a quantum wave pulsation.
It is the guiding force for gathering new informotion.

As I begin to live it with awareness, then I'll manifest a desirable new miraculous movie every moment! That is the purpose of this Decade: to encourage myself to live this wave quantum style. Magination gives me the ability to do this.

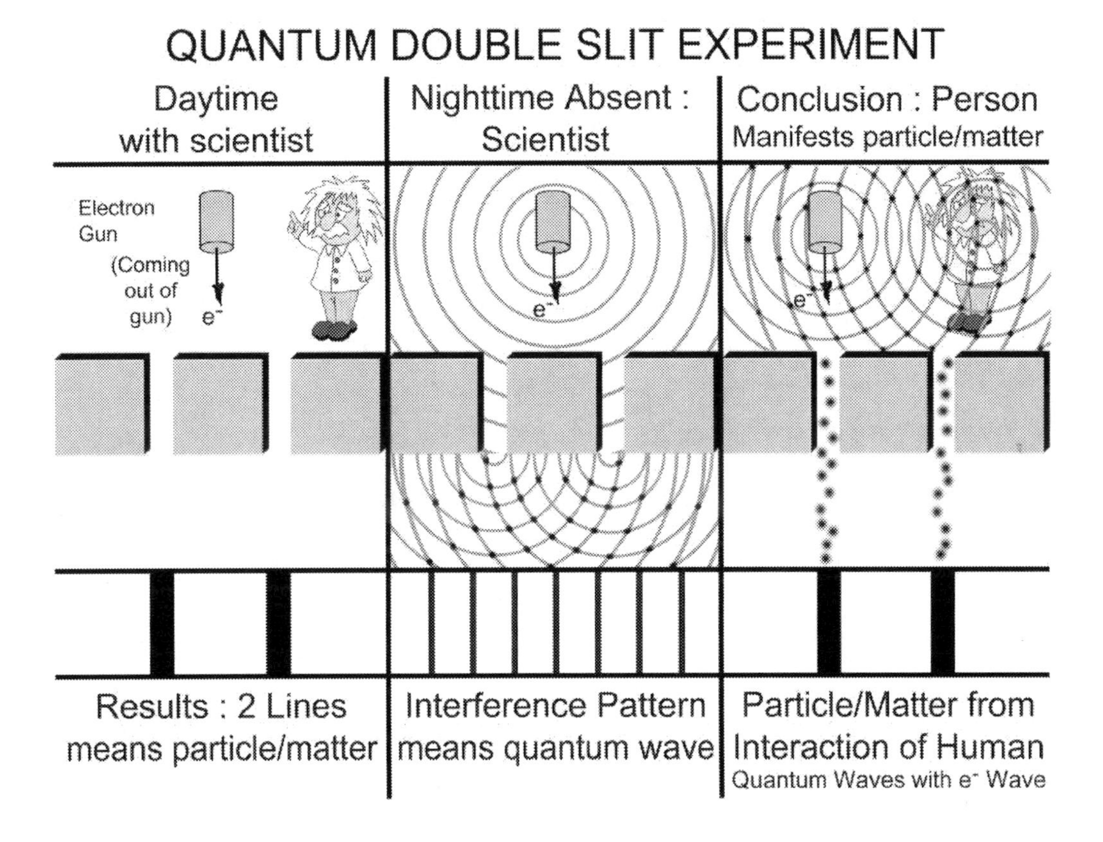

THE FAMOUS DOUBLE SLIT EXPERIMENT

This is the experiment that demonstrates that I manifest the universe.

Scientists designed the famous Double Slit experiment where they shot one electron at a bar with two open slits. They thought that if the electrons were balls they would go straight through those slits and hit a piece of photographic paper on the other side, leaving a mark. They discovered a picture of two lines on the photographic paper that matched the two slits. They concluded that the electrons were balls. To repeat the experiment they left the machine going while they went home to sleep.

On returning to the laboratory the next morning they developed the photographic paper. This time the photographic paper revealed hundreds of lines instead of the two lines. Hundreds of lines is called an "interference pattern" in physics. This means that the electrons are quantum waves, not balls. Please refer to the figure 3-1.

This upset them all including Albert Einstein. They could not understand how the electron could be a ball of matter and yet an invisible physical quantum wave at the same time. Did I mention that the quantum wave is as big as the universe?

After repeating this experiment many times they realized that it was their presence and actions that were manifesting the appearance of "spinning particles" from quantum waves.

Their conclusion is that we manifest matter. This really shook up all of their old beliefs.

There is a field of Quantum Biology, which has demonstrated quantum properties in biological systems. For further info watch Youtubes by Dr. Jim Al-Kahlili.

QUANTUM PLASTICITY

My body are like quantum silly putty; it transfigures as I live differently.

This has tremendous impact on my understanding of my body, living, and relationships.

I know now that the quantum spinning "particles" ARE really quantum waves turning around to radiate out, which structure the universe. This is called a quantum Wave Structure of Matter. Wolff, M. (2008), p. 7.

LIVING THIS QUANTUM UNDERSTANDING WITH EVERYTHING I DO IS MY KEY TO AWARENESS THAT EACH MOMENT IS AN EMERGENT MIRACLE. Let's 'begin again' in this moment.

DAY 1
WHO IS IN CHARGE OF LIVING MY LIFE?

MORNING PRACTICE: Say my Decade One quantum replacements and quantumcodebeing sentence once out loud with my eyes open.

I need to begin to imagine my living the way I desire according to our own homemade abidingself. When I forget this I fall back into cause-and-effect.

As my personal dream revealed last night: any focus on dismantling the old way is just like taking goat head stickers off of my body. Each removal was irritating, upsetting, and an annoyance. This dream showed me what I have been doing for years as witnessed by my patient and persistent coach Roger B. Cotting.

I have finally realized from my dreams, which according to quantum wave living IS my sacred teacher, that I need to exercise and emphasize the ability of 'magination' and direct perception.

Beginning to live without seeing problems or being upset is part of this big picture. Removing stickers is a bridge which needs to be traversed quickly as it is not quantum wave living.

Dreaming is quantum 'thinking' with the googleverse (universe). Communicating what I am doing in the moment, to give me the ability to replace it with my true desire options. Dreams as my sacred guide is coming up in the next Decade.

So I watch my day today and notice who is in charge of my living?

Am I in charge of manifesting my true desires as my abidingself? If not, I will do some creative work tonight to put me in charge? Later I will see how I give form to my true desires.

EVENING PRACTICE: Who is in charge?
I will write my responses to the following questions in my notebook.

1) ***Who did I decide was even partially in charge of my living***?

There may be someone that I am reacting to that bugs me? If this is true find the aspect of myself that they represent on the Personal Mirror-call Wheel Fig. 4, DECADE 4 and write it down to replace later.

Or, the one in charge might be someone I WISH were there with me; even if I don't know them yet. It may be an imagined boyfriend or girlfriend, a child or grandchild I wish I had OR HAVE, a family member, or a religious figure. I need to be honest with myself so I can benefit from this exercise. Write down all my ideas.

When muscle testing if I am not sure of who is in charge, I can become someone else I trust by asking, "Can I be my sister?" Usually the answer will be "yes". Then I ask as as my sister, about me. For instance, I would say, (asking as my sister), "Does Angela diminish her power when she is with her beloved?" Notice what answer comes into my awareness.* The answer is worth trusting.

If yes, I would face and embrace that, as a BATHWAVE waiting to be replaced:
Part 1: For example, "I face and embrace that I diminish my power when I am with my beloved." while touching lightly my whole body I speak it out loud.

2) ***Who would I like to be in charge of my living***?

When I am not in charge of my living, I am ready to replace this.

According to the airplane principle, if I don't put the oxygen mask on myself first, I can't help anyone else. This answer needs to be replaced with ME. Make a replacement to put ME in charge of my living.

Part 2: Put into my midline "I can, I will, I am in charge of my living, in relationship with my beloved."

I will add the present tense of this sentence to my list of quantum replacements.

3) ***Can I say goodbye to others who were being in charge of my living, with appreciation? YES!***

Just do it! Write it out so I can remember I am doing this.
It doesn't mean that they won't be there tomorrow or that someone I wish will be there. It means they won't be in charge of my living anymore. So they won't bother me or I won't need them to get through my day.

USE The BATHWAVE TECHNIQUE FROM THE FIRST Decade TO DO THIS.
PART 1. As I feel any other undesirable emotions about this person I face and embrace that. I feel these emotions and replace them with my desirable appreciating grateful emotions.

THE PART 2 replacement might be: I can, I will, I do appreciate my beloved for all the support I receive while I maintain responsibility for myself.

Add these to my quantum replacements.
Be patient with myself if I forget and regress.

4) ***Imagine a thank you sentence to you from the person who used to be in charge of your living.***

"I am thanking you for the opportunity to have been in charge in your living and I am glad you are living self empowered in this moment."

5) ***Notice how living might be freer and more creatively different tomorrow.***

Say all my replacements and abidingself before sleep.

*Each day I record my dreams. I am gathering information for my dreamwaving chapter. I do not need a whole dream. I just need a piece, either at the beginning or ending of the dream.

DAY 2
VISIONING MY DAY

MORNING PRACTICE: Say my quantum replacements and abidingself sentence out loud with my eyes open.

Vision my day being different now that I am in charge of my living. Write a couple of sentences about how I see my day being different today: even something as simple as not complaining about someone. Go ahead and be as creative as I feel like. Do automatic writing when nothing comes. That means write without editing or trying to make sense. Write on!

EVENING PRACTICE: INSTANTANEOUS VISIONING
The difference between visioning and visualization is that visioning is instantaneous!

This is why I need to practice visioning my true desires. I call this practice visioning with the understanding that as a vision it is true. In a quantum wave sort of way, the vision is immediate, invisible yet physical. I may not perceive the particle core of my quantum wave instantly, yet I know the wave is there in the universe and I carry on with living my true desire.

You step into it...

Waiting, doubting, or expecting results or proof is one form of self-sabotage.

This is where trust and belief enter in importance.

VISIONING EXERCISE OVER THE MOUNTAIN:

I need to read this slowly into a recording device or ask a friend to read to me in a dreamy voice as if I were walking, without any interruption of any sort. Phones off. (If no friend is available or recording device or voice memo on a phone just read it a couple of times till I see the picture, then do it from memory).

Once I am ready lay on a bed or floor relaxed yet attentive with eyes closed.

Turn it on and here I go:

I am walking on a path, on a sunny day, that is very easy for me.
The path begins through a large field of flowers that are my favorites. Look to the left and notice the colors that I like and look to the right and see some new flowers that I enjoy as far as my eyes can see.

Keep walking because up ahead I see a beautiful forest that the path goes through before I reach the mountain trail.

The trees are beginning to get bigger and thicker. I hear some birds singing. And begin to smell the fresh aroma of pine trees. The trail begins to climb and I know I am at the base of the small mountain that I am excited to get over it since I don't know what is on the other side. The path is safe and lined with large beautiful stones amid trees. I have never seen such large rocks before. I am enjoying climbing effortlessly full of energy. Up ahead I notice a small sign carved out of wood. "Tunnel ahead thru mountain". I am only half way up when the entrance to the tunnel appears.

As I approach it I feel relieved as I see lights lining the sides of the tunnel. I enter walking slowly and confidently. The path is wide enough for three or four people...and it's nice and cool inside. Walking almost a block of 100 meters (yards) I begin to see some sunlight at the end and I feel excited to discover what is out there. The closer I get I see a thick fog like cloud covering the exit. I walk slowly.

Standing in the cloud and looking down toward my feet I notice what is on my feet. Please remember what you see and share it.

Reader:
Stay silent for many minutes until they speak
After they say something, instruct them to open their eyes now. Ask them how they feel?
Tell them that they just had a brief experience of magination describing what they saw at their feet, no matter what they saw as long as they say something. They may say shoes, rocks, water, snow, or

fog; it doesn't make a difference. If they fell asleep you may do it again if time allows. Read a little faster and louder?

Being ready, willing and able to see what I desire is a gift of quantum wave living called magination.

Practice that as long as I can throughout my day tomorrow and even use magination picturing the kind of dream I might like to have tonight.

Say my quantum replacements and abidingself sentence once out loud with my eyes closed.

DAY 3
SEEING OTHER'S ABIDINGBEINGS

Whenever my personalself is living my abidingself, which is my quantum wave pattern, there is a sense of natural satisfaction.
There is a settling feeling in the gut. Magination means to see what I truly desire. Directing perception means living that without trying to cause it.

WHATEVER DOINGS FIT MY ABIDINGSELF's ACTIONS ARE MY AUTHENTIC LIVING. I HAVE A SENSE OF IT; ESPECIALLY WHEN I SEE OTHERS LIVING IT.

MORNING PRACTICE: Say my quantum replacements and abidingself sentence once out loud with my eyes open.

> "Practice seeing others as eternalbeings! …seeing myself with ordinary eyes keeps me from seeing myself as extraordinary …or eternal! …and seeing myself with extraordinary eyes that are normal for me …certainly keeps me from seeing myself as extraordinary! …in the same way, I cannot see or know myself as eternal while eternal! …therefore, seeing others as eternalbeings enables me to know and experience myself as eternal!
>
> Know myself and I will not actually know much about myself! …for knowing myself is too self-referential to be of any real value!... however, truly and genuinely knowing my neighbor as an eternalbeing …and you as a resplendent illuminatedbeing will become obvious!" Cotting, R. B. (2018)

EVENING PRACTICE:
Was I able to see others as an eternalbeing (quantum wave as vast as the universe)? Write what I observed. Read or repeat my quantum replacements and abidingself statement out loud with eyes closed.

DAY 4
MAGIC-NATION HAIKU

Vision it forward
As my magic-nation:
Initiates my self.

Directing perception is a lot like automatic writing! Another name for directing perception is MAGINATION or as a child called it 'magic-nation'. It is speaking and living authentically from my eternal quantum wave essence without editing.

Originally most languages were written and spoken as images or pictures like the characters of the oriental language or the hieroglyphics of Egypt. Writing and speaking in images encourages manifestation by direct perception. *DREAMS SPEAK THIS WAY SINCE THEY ONLY USE IMAGES!*

Western European/American languaging is structured in the old Newtonian physics of cause and effect. That is partly why it generates less artistic expression than people in Asia or ancient eras. It tears us apart into subjects and objects with causes and effects like judging with accusing, blaming, and complaining, ...any of the 8 basic bottomlines, all of which do not exist in the quantum world.

When I use words to describe an image, I expand my awareness manifesting emergent miracles!

When the image that is described is initiating my authentic desires, instead of judging 'good or evil' as a cause or effect, I untangle the entanglements of the world I do not desire!

I am supporting the true desires of my selves using this way of speaking in image descriptions. My true desires can then be supported by the universe, which can only mirror actions and pictures. As I actively live my true desires the universe can mirror that.

With all these tools, I am recreating my past physically as I desire, according to my vast abiding quantum wave self.

I live pictures of how I am eternally in my quantum wave's image by initiating "magination" of my true desires. My true desires are not things, but 'how' I do everything that I do. The doors and windows fly open to me! I am at home wherever I go.

I swim like a sea
Turtle self-illuminated
Home on my back!

MORNING PRACTICE: Say my abidingself and quantum replacements with eyes open. I spend this day using images or picture descriptions as my self-expression. This is called direct perception imaging in all my daily activities.

EVENING PRACTICE: Magination is different than meditation.

> "12...for my first foray into magination in search of direct perception, forget everything you've ever heard about sitting in mediation with my back straight, hands resting lightly and relaxed in my lap, eyes closed ...or focused on a point about 5 or 6 feet in front of you,....or concentration on my breathe, while calming my mind and letting thoughts come and go ...the purpose of magination is to bring direct perception into my life 24/7! ...therefore, this is not a "sitting exercise"! ...magination is to be a conscious part of everything you do ...not merely in a seated position! ...and certainly not as a meditative state! ...magination is not a separate activity! ...but do not focus on magination while driving or operating machinery or hand tools until you know how magination affects you!" "Magination" Cotting, R. B. (2018)

Dreams are examples of one type of magic-nation except I may not be doing it with awareness. Ask to remember that I am aware when I am dreaming tonight. Look at my hands in the dream to inform myself I succeeded. Do not worry if I don't as this is a life long practice called lucid dreaming. There are ways when awake tthat I discover direct perception.

Say my quantum replacements and abidingself sentence once out loud with my eyes closed.

DAY 5
"You are in everything around you! So look around"

MORNING PRACTICE: Say my abiding self and quantum replacements with eyes open. Practice seeing everyone as resplendent as I am.

> 47 ... "my declaration of how I will be eternally alive is a particular yet vast self-chosen blueprint of my destiny! ...the universe does not care for me as I should be! ...it functions to care for me as I truly desire!

> 49 ... contrary to common belief, the "real me" is not found in some secret or special center of myself! ...I see myself in everything around me! so look around, ...not within! ...and in looking around, no matter what I perceive, know that is as I am! ...seeing through personal·eternal abiding communion-eyes makes everything clear and understandable!" Cotting, R. B. (2018)

Today I perceive everyone resplendent. I see myself that way too living my true desires. I practice seeing everything I am doing as my desire. That might change how I feel about them? Check it out!

EVENING PRACTICE: DELIGHTED DOING NEW OPTIONS OF WHAT I TRULY DESIRE?

> 50 ... what do I truly desire? ...I'll venture to declare that I want to be personally alive with matters of interest, with wisdom, and with robust wellbeing! ...I want to be assured ...and feel secure in everything I do, you want awareness of a true purpose and meaning to guide me! ...and I want to be genuinely affirmed. Cotting, R. B. (2018)

Write about my understanding of how living my eternalself quantum wave pattern gives me all this. The universe has to affirm it by entanglement in each moment!

Am I perceiving this or missing the obvious!? It's ok if you miss this.

Magination is my Magic-nation.
Magic-nation is where I live with quantum wave living.

Say my quantum replacements and abidingself sentence once out loud with my eyes closed.

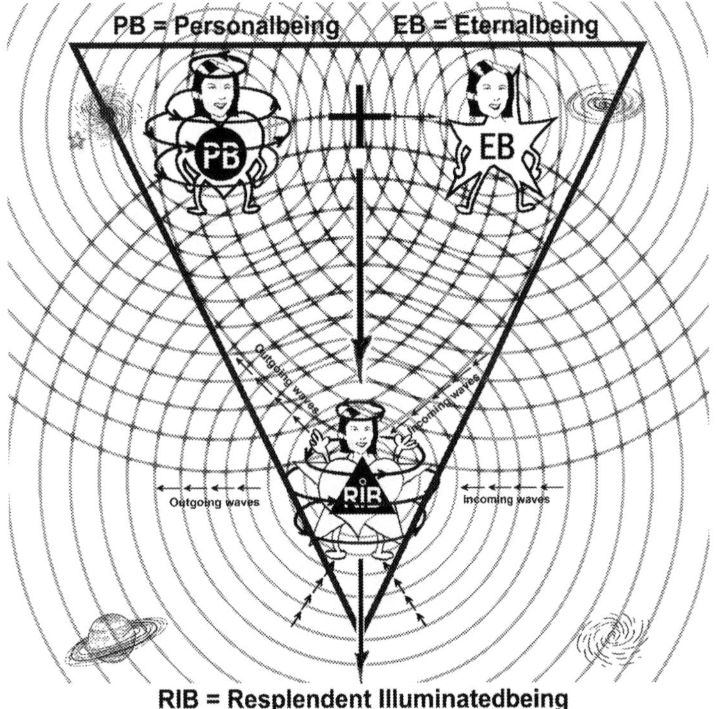

DAY 6
MAGINATION REQUIRES A FOCUS AND AN INTEREST OR A QUESTION

MORNING PRACTICE: Repeat quantum replacements out loud eyes open.
Record my dream before I turn my light on. I can use a recording device like a voice memo or write in the dark. I can prepare myself with paper and a pen near my bedside.

Direct Perception of an Image of My Eternalself:

> "(13)... magination requires a focus and an interest or question ...for direct perception to respond to you and for you to understand the response! ...a fine beginning focus is to create an image of my self and life as an eternalbeing! ...as an eternal physical being ...not as a spiritual being that is vague and unknown and unknowable! ...
>
> EXAMPLE: the image I created of myself as an eternal physicalbeing is as a teacher and advisor in the Parthenon in Athens! ...I chose an image from this time period to give my eternalself a past that has always been with me so I can feel and understand how my life today has developed along the lines of this eternalself and as a new

> beginning (an incipit) for every aspect of my self and life! ...such that, relative to that eternalself I know my life today and for the future! ...to come alive for you ...my eternalself needs a desirable past! ...so I can give my eternalself and me a future that follows!
>
> (14) ... what image do I have of my eternalself? ...am I living that image in my life today?
> Take a few moments and create that image! ...don't wait! ...and don't let thinking or reasoning talk you out of doing this! ...what is my eternal name? ...and what am I doing? ...am I willing to create every detail of my eternalself? ...if so, every time I enter magination, add more and more details to my eternalself and life!" Cotting, R. B. (2018)

Write my answers to Rogers suggestions above. I can do automatic writing if nothing comes to mind. Don't worry if it doesn't make sense now. Whenever I have a free moment today think of my image of my eternal self. Communicate my eternalself through my interactions with people today.
Feel some appreciation of my resplendence thanks to my eternal self. Add details to my eternalself if I feel so moved.

EVENING PRACTICE:
Write about my interaction with others and myself today. Did my day seem different than previous days? Give some appreciation of my resplendence thanks to my eternalself. Add details to it if I am so moved.

Ask for a dream or inspiration to add to my image of my eternal self tonight.
Write about it immediately upon awakening.
Say my abidingself and quantum replacements with eyes closed.

Communion of PB : EB
Is Resplendent Illuminatedbeing

DAY 7
The Universe is Good And Benevolent. How?

> "15 ... Newton's third law of motion states, that for every action, there is an equal and opposite reaction ...this means: the universe only reacts! ...and even reacts to its own reactions! ...which has led to the mistaken, belief that the universe is conscious ...and acts intentionally!...actually, the universe only reacts!...and is reacting to you all the time! ...the universe lives my life moment by moment ...right along with you, ...and stores that memory in matter and in space for you to continue living or struggle against!" Cotting, R. B. (2018)

This is true because of the quantum secret of entanglement spoken of in Decade 1.

MORNING PRACTICE: Say my quantum replacements and abidingself sentence once out loud with my eyes open.
Discuss something that happened yesterday that really was something I asked for. Take note about a reaction of the universe supporting me.
Today say and do those things I would like to have supported by the universe.

EVENING PRACTICE:
How did I do today? Was today different for me to say and do what I desire the universe to support? Say replacements out loud with eyes closed and your abidingself quantum wave pattern.

DAY 8
Seeing Options Without Judging

MORNING PRACTICE: Say my quantum replacements and abidingself sentence once out loud with my eyes open.
Write dreams down with no judging. If I cannot remember chose a daily event from yesterday and write it as if it were a dream with no judging. It may take two attempts.
This is an important practice of observing and a first step in any nonviolent communication*.

Today I practice observing activities with no judging just observe what is happening.

Subtle examples of judging, for example like using superlatives, this is better to do, the best way, most words you add est to. Comparisons are judgements, using good better or best ideas. Judging gets very subtle. Accusing, blaming, or complaining plus all the eight bottom lines are all forms of judging. Should and ought are judgements.
Practice listening to connect to others feelings and needs with people I interact with. I can watch Marshall Rosenberg, author of nonviolent communication method on YouTube**. For example, I can say; did I hear that you are feeling ...because you have a need of ...? When they agree then say, Have you considered this possible option ...? This is NVC's method simplified, instead of using judging language. Or read quantum communication to practice with a friend willing to learn it. Take note of a few examples of moments I observed, to use in the evening exercise.**
**For some inspiration watch Marshal Rosenberg videos on NVC (Non-violent Communication) on YouTube.

EVENING EXERCISE:
Write an example of an observation I witnessed today. Without judging, write what I would change it to, if I had any choice possible. Call this an option without judging.

Now do it about tomorrow and pick something I do everyday. Write how I see myself doing it in a new way. This is direct perception practice. The more I picture myself doing what I really wish the more it occurs along with my eternalself action.

Say my quantum replacements and abidingself sentence once out loud with my eyes closed.

In preparing for the Pregame for Decade 4 I will need to gather the following materials:

- Seven paper bags, boxes, or other non transparent containers that are the same
- Uniformed paper - same size, same shape, same texture, same color
    - 13 sheets of printer paper work well just fold into fourths (by folding in half and then half again) and cut or:

    - 52 smaller squares/rectangles
- Marker for writing on bags
- Pen/pencil for writing on paper

DAY 9

"Would I be willing to do or say this eternally and have others do or say this eternally to me?"

MORNING EXERCISE: Say my quantum replacements and abidingself sentence once out loud with my eyes open.

> "consider following this process: ...in my daily life, before you act or speak choose what you will do or say! ...then ask myself, "...would I as an eternalbeing be willing to do or say this eternally and have others do or say this eternally to me? ...if my answer is "yes" then do or say as you have chosen, but if the answer is "no! ...I wouldn't do or say this eternally or want others to do or say this to me eternally", then make another choice of actions or words!
>
> This may seem overly complicated, yet, consider that in living as an eternalbeing you would always be doing and saying only what you would be willing to express and have expressed to you eternally! ...after a while, as my choices of doing or saying are as you are willing to act and speak eternally, you can stop questioning and confidently live my choices as eternalbeing-choices!" Cotting, R. B. (2018)

Ask myself this question about my interactions today:

Would I be willing to do or say this eternally and have others do or say this eternally to me?

EVENING EXERCISE:

> "Good/evil or right/wrong judgments are not necessary in an eternal life! ...nor should they even be considered! ...for all such judgments are languaging ...not directing perception! ...making this choice is direct perception! ...and acting upon that choice makes it real in my body and life ...and in the world around you! ...and in that you offer a judgment-free space to communion with you! ...and form an awareness of myself as an eternal abidingbeing and experience myself in that way!" Cotting, R. B. (2018)

There is no judging of me in my dreams. My quantumself is showing myself what I am doing to have options if I desire them. Ask to remember a dream tonight.

Say my quantum replacements and abidingself sentence once out loud with my eyes closed.

DAY 10
NOTHING TO FIX

Directing perceiving means to see what you truly desire.
Think about how much of my day is spent; worrying, coping, struggling, fixing, stressing, judging, etc. We call this "languaging".

> "48 ... in my constant struggle to cope with presumed harm and difficulties ...and seeming limitations, instead of resolving these problems, my beliefs and feelings take on the confusion, distortions, and burdens of language, and my perceiving-eyes are blinded and deceived by languaging!" Cotting, R. B. (2018)

MORNING PRACTICE: Say my quantum replacements and abidingself sentence once out loud with my eyes open.
Today catch myself doing any related actions listed above and/or the following: confusing, distorting, deceiving, blinding, fixing, struggling, stressing, and so on.
Then I can replace it with what I truly desire to be doing: relaxing, clarifying, respecting, understanding, trusting, being honest, confident, secure as my eternalself.

Then there may be nothing to fix?

EVENING PRACTICE:
"We don't see things as they are. We see things as we are." Anais Nin. This is a poetic expression of quantum entanglement!

Write about something I remember seeing today that I liked and guess what it might mean about how I am!?

Before sleep do your quantum wave pattern and replacings out loud with eyes closed.

I pat myself on the back for completing yet another Decade of quantum universal thinking.

OPTIONAL METHOD CALLED GAME to use in DECADE 4:

Googleverse Access Method Express: GAME

Even if I am not an experienced muscle tester or have another method to access informotion I recommend using this GAME when feeling dehydrated, blocked, stressed or confused.
This will take 30 minutes to an hour to create, so find your favorite playlist and enjoy! It is time to introduce myself to a game that can help me access informotion from the quantum memory of the

universe. This technique can be used if muscle testing and other methods do not work for me. It is important for me to follow these directions and create a uniform tool to use. If my tool is not uniform the results may be skewed.

Materials Needed:

- Seven paper bags, boxes, or other non transparent containers that are the same
- Uniform paper - same size, same shape, same texture, same color
    - 13 sheets of printer paper work well just fold into fourths (fold in half and then half again) and cut, or
    - 52 smaller squares/rectangles
- Marker for writing on bags
- Pen/pencil for writing on paper

Follow these steps to create an Access My Googleverse tool!

Step 1: Label each bag

1. Yes or No
2. Six Areas of My Life
3. Five Elements
4. BATHWAVEs
5. My Personalself Aspects
6. Ages or just multiples of 10
7. 1-9 Numbers (to be Ages, Days, Months, Years, or Grades in School)

Step 2: Fold 13 letter/legal size sheets pieces of paper in fourths and then cut. I now have 52 uniform pieces of paper.

Step 3:

Contents for Bag 1: Yes or No

- Write Yes on 4 of my pieces of paper.
- Write No on 4 of my pieces of paper.
- Fold each paper in half 3x yes and no inside.
- Place all 8 pieces of paper in the Yes or No bag.

Contents for Bag 2: Six Areas of My Life

- Using 6 pieces of paper I will write one area and options on each piece.
    - Family: mother, father, brother or sister, School: teacher, students, grades?, Death or Sickness, Moving My Location, Career or Job, Religion or culture
    - If I choose one with options, I will use Yes or No bag to select.
- Fold each paper in half 3x.
- Place all 6 folded pieces of paper in the Six Areas of My Life bag

Contents for Bag 3: Five Elements

- Using 5 pieces of paper I will write one element on each piece.
    - Metal, Water, Wood, Fire, Earth
- Fold each paper in half 3x.

- Place all 5 folded pieces of paper in the Five Elements bag.

Contents for Bag 4: BATHWAVEs

- Using 9 pieces of paper I will write one BATHWAVE on each piece
    - Beliefs, Actions, Thought, Habits, Words, Attitudes, Values, Emotions, Judging
- Fold each paper in half 3x.
- Place all 9 folded pieces of paper in the BATHWAVEs bag.

Contents for Bag 5: My Personalself Aspects

- Using 6 pieces of paper I will write one aspect of my being on each piece.
- Father=Self-guidance, Mother=Self-nurturance, Female Child=my Creativity, Male Child=my Creativity, Female Beloved=self love, Male Beloved=self love
- Fold each paper in half 3x.
- Place all 6 folded pieces of paper in the My Aspects bag.

Contents for Bag 6: Decades called multiples of 10 (Ages, days, months, years)

- Using 10 pieces of paper I will write one Decade on each piece.
    - 10, 20, 30, 40, 50, 60, 70, 80, 90, 100
- Fold each paper
- Place all 10 folded pieces of paper in the Ages bag.

Contents for Bag 7: Numbers (Days, Months, Years, or School Grades)

- Using 10 pieces of paper I will write one number on each piece.
    - 1, 2, 3, 4, 5, 6, 7, 8, 9
- Fold each paper in half 3x.
- Place all 9 folded pieces of paper in the Numbers bag.

How This Works:

- Ask my question to a specific bag; for example, "Five Elements, which element will help me ______?" or "Yes - No, Is this about my dog?"
- Shake the bag before selecting a slip of paper.
- Reach in and pull out 1 slip of paper. Pull out the slip of paper I gravitate towards. It may be the first one I touch or I may have to dig around a bit.
- Open and read response.
- Refold the paper the **same way** and put the slip back into the bag.
- Ask another question if more information is needed, repeating this process.

Tip: Using the Decades/Multiples of 10 and numbers bags together you can receive an exact age. For example if I pull out a 30, I know my answer is in your 30s. I can then ask if I was 30 year of age. If I get a no, I can ask what age in my 30s? Then pull a number from the Numbers bag to receive my answer. I pull out a 4 meaning I am 34 years old.

Now I am ready to Access Googleverse!

Read a clients experience using the Googleverse Access Method Express (GAME):

In order to put this tool in the book we had to run an experiment. One of the quantum life coaches tested out the Googleverse Access Method Express tool to see if accurate results were received. She asked a question and received answers by pulling slips of paper from the bags first. Then she used muscle testing to check for accuracy. She writes about her experience here:

Throughout my day I was aware that a pattern was returning and I had not been able to shift it. I knew the pattern had something to do with trust yet I could not place it. I thought I would try out the GAME tool to see what results I would receive.

Ask a question, receive an answer…

"Yes-No (bag): can I ask about my trust pattern?"
Bag - Yes Muscle Test - Yes
"Yes-No: do I use the Five Elements bag?"
Bag - Yes Muscle Test - Yes
"Five Elements, which element will help me understand my trust pattern?"
Bag - Fire Muscle Test - Fire
"BATHWAVEs: which Fire BATHWAVE is related to my trust pattern?"
Bag - Habits Muscle Test - Habits
"Yes-No, which Fire organ habit, heart?"
Bag - Yes Muscle Test - Yes
"Yes-No, is there another Fire organ habit other than the heart?"
Bag - No Muscle Test - No
"Yes-No, is it heart habit 1 (referring to the heart BATHWAVEs chart)?"
Bag - No Muscle Test - No
"Yes-No, is it heart habit 2?
Bag - Yes Muscle Test - Yes
"Yes-No, is it heart habit 3?
Bag - No Muscle Test - No

Part 1 **BATHWAVES** I Face and Embrace... in whole body	Part 2 **REPLACINGS** around my Midline with Hands, I Replace with Grace. I can, I will and I...
I face and embrace my love letter that I deny my entinglement mirror-calls.	I can, I will and I am replacing my love letter with appreciating the power of my entinglement.

"Yes-No, am I replacing my love letter with appreciating the power of my entinglement?"
Bag - Yes Muscle Test - Yes
"Yes-No, am I replacing my love letter with appreciating the power of my entinglement at 5 years?"
Bag - Yes Muscle Test - Yes

"Yes-No, am I replacing my love letter with appreciating the power of my entinglement at 10 years?"

Bag - No Muscle Test - No

"Yes-No, will the Decades bag help me?

Bag - Yes Muscle Test - Yes

Decade, at what age did this trust pattern begin?

Bag - 70 Muscle Test 70

"Yes-No, is it the age of 70 months?

Bag - Yes Muscle Test - Yes

"Yes-No, will the My Aspects bag help me understand my trust pattern?

Bag - Yes Muscle Test - Yes

"My Aspects, which aspect of me has an issue with trust?"

Bag - Female Beloved Muscle Test - Female Beloved

"Yes-No, do I have enough information to replace this pattern?"

Bag - No Muscle Test - No

You will learn this relationship method in the next Decade.

I continued asking questions until I was able to access enough information. What I discovered was that I stopped trusting my female beloved sister around the age of 6.

"I face and embrace that I stopped trusting my female beloved when she left me. I felt sad and neglected and I am ready to replace this.

"My female self-love can, will, and is gentle and warm with my creativity."

"Yes-No, did I replace this pattern?"

Bag - No Muscle Test - No

After asking a few more questions I began to have a mini tantrum during this game. I was on the verge of feeling quite frustrated when I received a call from Dr. Longo. What we discovered was this pattern also occurred at school with a teacher around the age of 6 as well as occurring in the womb. I felt cold and disconnected, even in the womb of my mother. We made several shifts starting with not wanting to be visible, feeling cold and disconnected from both my self-nurturance (Mom) and myself, not liking myself, and not liking my warm interactions with others! I felt greatly relieved after these replacements!

Dr. Longo reminded me that Living Quantum Resplendence is a treasure hunt. I am on a treasure hunt, following the treasure. Sometimes I may get misguided, lost, or even discouraged yet the treasure will always be in sight. I may even have to find the same treasure a few times before it really sticks. Years of old habits may need some additional attention. :-)

EXTRA DAY TO PRACTICE USING THE GAME IF YOU LIKE

MORNING PRACTICE: Say my quantum replacements and abidingself sentence once out loud with my eyes open.

Pick a dream or daily event, or anything, that I do not understand and would like more clarity on. Ask which of the Five Elements will help me to GAIN CLARITY. When I get one or two elements ask which of the BATHWAVES is pertinent with the first element. Look up the element Decade and the corresponding chart with the BATHWAVEs. After choosing or muscle testing the appropriate BATHWAVE I will **face and embrace** to **replace it**. After completing the replacement, see if I have gained clarity.

EVENING PRACTICE:
If I would like more clarification on the same dream or event or I can choose a new one to apply the Six Areas of Living. Go back to morning practice of Day 2 of this Decade, if needed. Using my chosen method for testing, select one or two of the six areas of your life and continue to ask questions. I may ask about my age to discover the issue that is relevant to that area of my life in relation to my dream or daily event. Keep following the treasure hunt until an understanding appears.
Now I have a good overview for relating to and freeing up my living.

FINDING THE ANSWERS TO ANY QUESTION USING 6 AREAS OF LIVING AND/OR 5 ELEMENTS, AS WE DID IN GAME

When using muscle testing, pendulums, or any other method please refer to this outline to access any informotion.
SIX AREAS OF MY LIFE include the following: family, school, job/career, death/sickness, moving my location, and belief systems.
First line of questioning: The Six Areas of My Life:

1. Is it family? If yes, delve deeper by asking:
    a. Is it about parents: mother or father?
    b. Is it about siblings: brother or sister?
    c. Grandparents: which one?
    d. Spouse?
    e. My children: which one?
2. Is it about schooling? If yes, delve deeper.
    f. Teacher?
    g. Fellow students?
    h. A subject or grades in school?
3. Is it about job or career?
    i. Boss?
    j. Colleagues?
4. Is it about a death or a sickness?
5. Is it about moving my location?
6. Is it about belief systems: society, culture, gender or religion?

The second line of questioning uses the **FIVE ELEMENT PROCESS** that I have developed and their BATHWAVEs.
Step 1): Choose an element using my method: fire, earth metal, water, or wood.

Step 2): When using the GAME choose a single number from the numbers bag (must be less than 8). Choosing the amount of BATHWAVES (from the BATHWAVE bag) indicated by the number drawn from the numbers bag.
Step 3): Access the element BATHWAVE charts which can be found in Decades 7-14 or in the appendix.
Step 4): Do my replacing and write the present tense replacements down in my notebook.
In this work, we do not look for stories. We look for the **verb** or (**action**) and occasionally the **feeling** (**adverb**), or a combination of these two. We then use the mirror-call and Ace of BATHWAVE techniques.

In order to Shift Relationships We Invent the Mirror-call Parts of your Personalbeing:

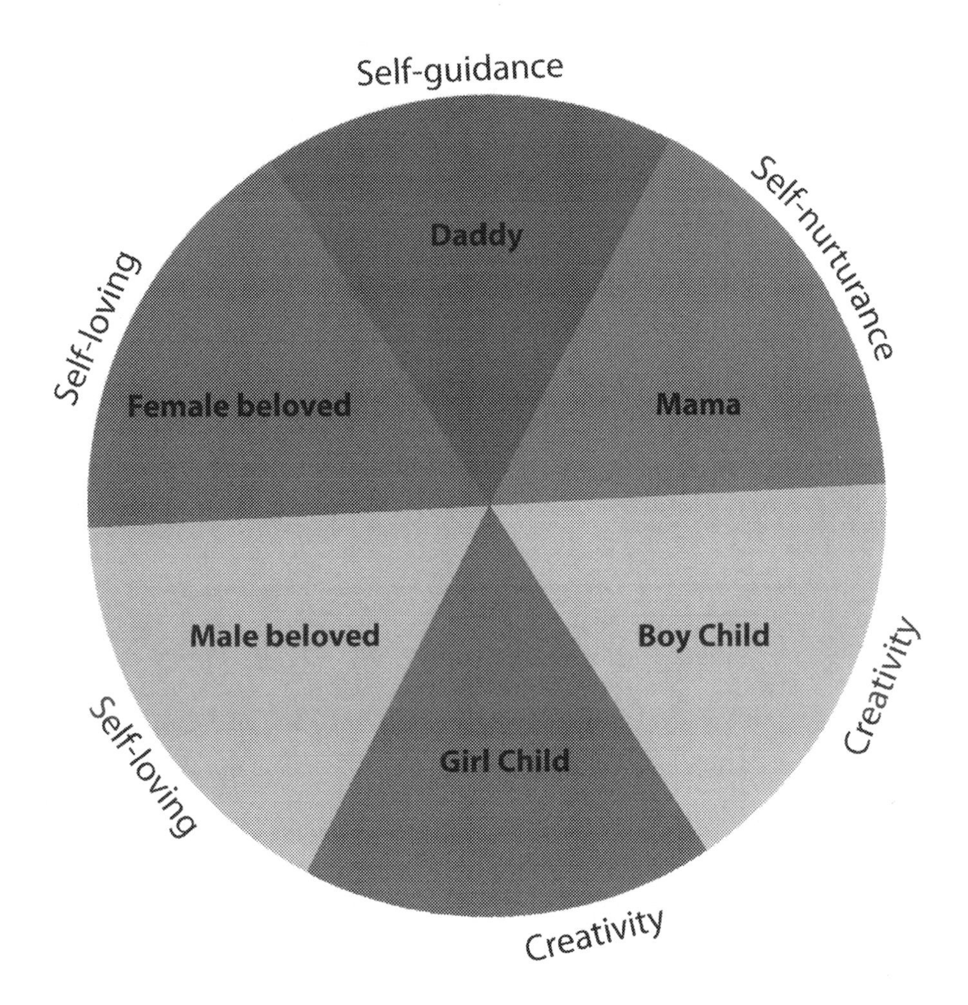

Inside the circle find the person who is in the shiftable relationship...
Outside the circle read the shiftable mirror - call part of yourself...

Method:

1. Speak to a mirror representing the person you are in relationship with.

2. Write down what you wish the person would change about themself.
Find the piece of the pie that is their relationship to you.

3. Look up an the outside of the circle the mirror call part of yourself that they mirror for your illumination.

Table 4

DECADE 4

REPLACING OLD PATTERNS IN RELATIONSHIPS: RELATIONSHIFTING MIRROR-CALL METHOD (M&M's)

MIRROR-CALL RELATIONSHIPS HAIKU

or

ENTINGLEMENT SPEAKS

Looking directly
Into your eyes as a mirror:
I speak to self.

DAY 1

Relationshifting: REPLACING YOUR STARTER BODY'S PAST

A Dutch psychologist and extraordinaire co-facilitator of family constellation work that I exchanged with in Thailand at Tao Garden Health Resort wrote to me later:

> "I feel my family constellations method is very close to the quantum approach. Every moment we create our own experience in the world. Outside events are a perfect mirror of our mind-state and are only there to help us when we look at them with love." Dr. Marita Schropp, Dutch psychologist (2018)

She enters into entangelement with her characters, then acts out their essence to help you transform yourself. We use the mirror-call method a similar way.

MORNING PRACTICE: Say my quantum replacements and abidingself sentence once out loud with my eyes open.

U. S. A.

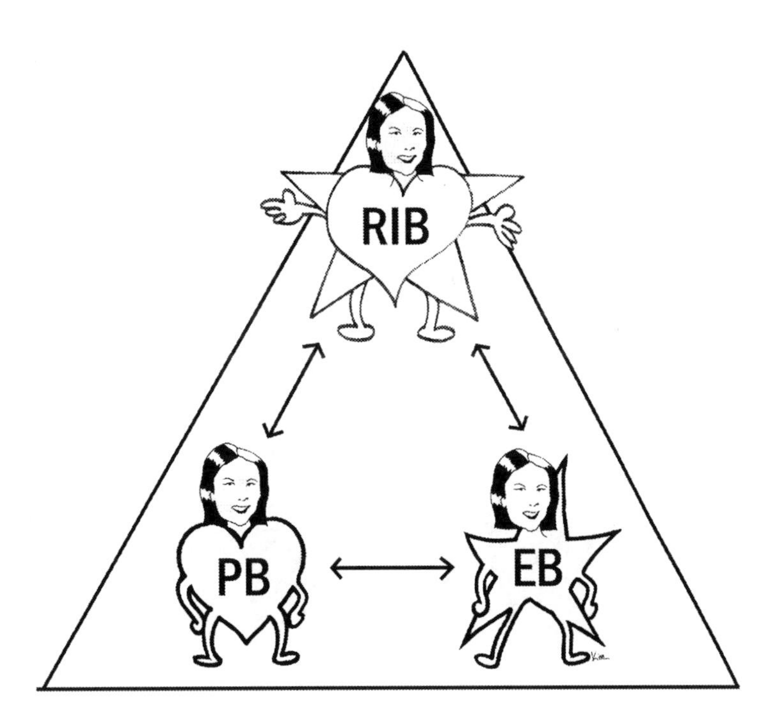

The Resplendent Triunity of You

United States of Awareness

MY PRIMARY RELATIONSHIP

Do I see that I have two natures with equivalents in the quantum world? The first nature is my particle aspect of me known as the personalself (my body/brain/personality). Historically this aspect used to live by "observation of what is happening to me" and 'cause and effect'. The second nature of me is equivalent to the wave of the quantum world and is called the abidingself. My abidingself lives by "visioning" what I desire in my living outside of cause and effect!

When my personalself communions with my quantum wave pattern all surprises break loose and miracles are afoot as I desire! THIS IS MY PRIMARY RELATIONSHIP.

Since this Decade is about relationships and my primary relationship is between my eternal abidingself (quantum wave self) and my personalself it would be good to get an image in my mind of me as my abidingself.

Jot down some ideas today until an image formulates itself. Picture what I would look like doing what I truly desire. Once, I saw my abidingself as a Native American weaver at a loom. Another time, she was in Egypt teaching. I have lived with the Hopi Indians and I do plan to meet an Egyptian coach there next year to share.

Mirror-Calls of my Resplendence

Mother - Son/Daughter

His/Her BATHWAVE to face & embrace:

I must still need a mother...nurturance

His/Her Quantum Shift:

I can give my own self-nuturance
I will give myself self-nurturance
I am self nutured

EVENING PRACTICE: MOM, MAMA, MOTHER - SELF-NURTURANCE

My mom mirrors my self-nurturance!

Let's use the mirror-call of relationships to transform our past and ease our personal starter body life. How I am reacting to a relationship is the love letter I give myself. To do this process use any handheld or wall mirror. Have paper and pen handy to write down what I say. If no mirror is available, I can talk to the palm of my hand imagining it is a mirror.

Here is the process:

I talk to my mother in a mirror. Tell her what I wish had been different between us like: "Mom, I wish you had appreciated me as I am."

The mirror frame is my mother, but who am I talking to in the mirror? The obvious answer is myself.

I ask myself <u>what aspect of me does my mother represent</u>? Look at the figure 4.1 to answer the previous question. Around the circle, next to mom we read <u>SELF-NURTURANCE. In this relationship I am a child, which represents my CREATIVITY.</u>

So here is the BATHWAVE of the mirror-call:

BATHWAVE OF THE MIRROR-CALL: Self-nurturance (Mom)	
BATHWAVEs Part 1: Face & Embrace	**Quantum Replacements Part 2: Replace with Grace**
I face and embrace that my self-nurturance does not appreciate my creativity as it is and I am ready to replace this.	And the replacement would be about how "my <u>mother</u> appreciates me, her <u>child</u>: Again, using my mirror-calls aspects on circle above, it sounds like this: My <u>self-nurturance</u> can, will, and does appreciate my creativity as it is.

Does that make sense? Don't worry if it doesn't make sense now; it will when I begin nurturing my creativity as it is!

When I am replacing the BATHWAVE move hands around my midline while saying out loud: **<u>My self-nurturance can, will, and does APPRECIATE my creativity as it is.</u>**

Write the present tense only in my journal to be repeated with all quantum shifts from Decade 1 for 21 days to 3 months until they feel natural.

AWARENESS THROUGH MOTION (ATM: There is lots of energy here ($))

This is all an example of seeing entanglement for my illumination.
Explanation of the entinglement (entanglement) : The universe only shows me what I am doing and this is one gift of entinglement. My response is mirroring me. I can access all the informotion as I ask for it.
Mother mirrors my **self-nurturance**. I am my '**creativity**' in the scenario with her.

Say my quantum replacements and abidingself sentence once out loud with my eyes closed. Remember to add my new replacements to my list in present tense.

DAY 2
KNOT ME? YES YOU, COULDN'T BE! THEN WHO?

MORNING PRACTICE: Give attention to dreamwaving. First, say my quantum replacements eyes open.

Visioning is new to me in that it is really a catalyst of my life. A catalyst does not do anything, a catalyst by its very presence activates and stimulates a phenomenon without moving, causing, or affecting, or being affected. It remains the same before and after an action is complete! That's how my abidingself works as a catalyst.

Today I think of gratitude to offer my best friend, my quantum wave self, by name. Extending my gratitude for visioning my six accomplishments and anything else I would like say to them. If you have resistance to gratitude, write down and embrace it on day 10 of this Decade.

Say my quantum replacements and abidingself sentence once out loud with my eyes open.

Mirror-calls of my Resplendence

Father - Son/Daughter

His/Her BATHWAVE to face and embrace:

I face and embrace that I still need a dad.

His/Her Quantum Shift to face and embrace:

I can give myself my own guidance
I will give myself my own self-guidance
I do give myself my own self-guidance

EVENING PRACTICE: PAPA, DAD, FATHER - SELF-GUIDANCE

My papa mirrors my self-guidance!

Speak with my father in a mirror. Say what I wish he would have been doing or not doing for me. Then write it down in journal.

Example: Dad, I wish you had not CRITICIZED me as a child.

Replace **father** with **self-guidance** and me as his **child** (no matter the age) with **creativity**.

Example:

BATHWAVE OF THE MIRROR-CALL: Self-Guidance (Papa)	
BATHWAVEs **Part 1: Face & Embrace**	**Quantum Replacements** **Part 2: Replace with Grace**
I face and embrace that my **self guidance** CRITICIZES my **creativity** and I desire to replace this now.	My **self guidance** can, will and does **CONNECT** AND **COMMUNICATE supporting** my **creativity.**

I look under complains/criticizes on 8 bottom line mirror-call behavior chart to find replacement. Write only the present tense of the replacement down in my quantum notebook to repeat for 21 days to 3 months.

Say my quantum replacements and abidingself sentence once out loud with my eyes closed.

DAY 3:
OUR QUANTUM WAVE PATTERN IS OUR CATALYST

MORNING PRACTICE: Say my quantum replacements and abidingself sentence once out loud with my eyes open.

The abidingself is the catalyst of my life. Without awareness of my abidingself, I have less power and ability to be who I really am. My abidingself does not do anything and it is never changed; it merely catalyzes my own emergent expansion!

Do automatic writing asking my abidingself by name to help me catalyze something specific I would like in my life. Then see myself doing what I asked for, as it is possible, picture or do it today!?

His BATHWAVE to face and Embrace:
My feminine side is not hearing my masculine needs and desires.

His Quantum Shift to replace with his grace:
My own feminine side can hear and support my needs and desires.
My own feminine side will hear and support my needs and desires.
My own feminie side hears and supports my needs and desires.

EVENING PRACTICE:

THE COOKIE JAR JINGLE HAIKU

Who took the cookie
From the cookie jar? Who me?
Yes you! It could be.

My **beloved** mirrors my **self-love** (in this case, my love of my cookies…lol). See fig. 4.3.

I speak into a mirror talking to my beloved even if I don't have one. Imagine one I desire or pick a past one or a wishful friend; they don't have to be alive. **Say what I wish he/she would have been doing or not doing for me**.

Example:

BATHWAVE OF THE MIRROR-CALL: Self-love (Beloved)	
BATHWAVEs **Part 1: Face & Embrace**	**Quantum Replacements** **Part 2: Replace with Grace**
I face and embrace that I wish my **self-love** was new and different each moment appreciating and cherishing me passionately.	My **self-love** can, will and is new and different each moment appreciating and cherishing me passionately.

Do my own BATHWAVE technique regarding my self-love (beloved) and add it to my quantum replacements which I say before sleep.

Say my quantum replacements and abidingself sentence once out loud with my eyes closed.

DAY 4:
TRANSFORMING MY CREATIVE EXPANDING FOCUS...

MORNING PRACTICE: Say my quantum replacements and abidingself sentence once out loud with my eyes open.

Visioning is about who I already am, for as soon as I vision, it is. This is an analogous to the quantum world phenomena called entanglement, so don't be surprised if you feel confused by entanglement. Einstein and all the physicists were so confused that Einstein called this spooky action at a distance. Using entanglement with vision would be pulling in a line throughout the universe instantly.

EVENING PRACTICE:
The **child** is a mirror of my **creativity**. Today I chose the 'masculine' child aspect of my living, which mirrors my active creative expanding focus.

Speak to a masculine child I have an issue with or to myself as I was as a child with my masculine aspects such as logical creativity, math, science, linear thought, athletic strength, etc.. Say what I wish was different. Write it down.

BATHWAVE OF THE MIRROR-CALL: Creativity ('Masculine' child)	
BATHWAVEs **Part 1: Face & Embrace**	**Quantum Replacements** **Part 2: Replace with Grace**
Face and embrace my creativity that is not active, expressive, expanding and focusing and I replace this.	Replace this with what I desire about my creativity: My male creativity can, wiil, and is active, expressive, expanding and focusing.

Add this quantum replacement to this Decade's list. Say my quantum replacements and abidingself sentence once out loud with my eyes closed.

DAY 5
TRANSFORMING MY CREATIVE OPEN SACRED SPACE

MORNING PRACTICE: After recording dreams, say my quantum replacements and abidingself sentence once out loud with my eyes open. Take my quantum vitamins!

Mozart explained that he would hear a symphony in an instant and it would take him long periods of time to write it down. This would be an illustration of resplendent, creative living. The vision as a quantum wave is the catalyst for the personalself, particle equivalent of us in the quantum world, which enables actions. The quantum wave doesn't cause the creative actions, it is only catalyzing like a vitamin does. The quantum wave doesn't change and it doesn't do anything; however, it is magical and our awareness of it really helps.

Today I notice my quantum wave presence as my catalyst, my quantum vitamin.

EVENING PRACTICE:
The **female child** is a mirror of my **open sacred space of creativity**.

Speak to a feminine child I have an issue with or to myself when I was as a child with my feminine aspects like spacial creativity, musical, literary skills, nurturing, emotional, ect.. Say what I wish were different. Write it down.

BATHWAVE OF THE MIRROR-CALL: Creativity (female child)	
BATHWAVEs **Part 1: Face & Embrace**	**Quantum Replacements** **Part 2: Replace with Grace**
Face and embrace this about my creativity is not opening sacred space.	Replace this with what I desire about my creativity: My female creativity. can, wiil, and does open sacred space.

Add this quantum replacement to this Decade's list and say my quantum replacements and abidingself sentence once out loud with my eyes closed.

DAY 6
BEING MY SELF-AUTHORITY

MORNING PRACTICE: Say my quantum replacements and abidingself sentence once out loud with my eyes open.

Intention is different from visioning. Intention is what I wish to be in the future whereas visioning is what I am in the moment. Visioning then creates observation. When I am living vision communioned with the body this is the secret of resplendent living. This is the living of the wholeness of myselves whose equivalent in the quantum world is the wavicle, so I am my visions!

Practice visioning my day as I desire. When possible I think of my quantum wave-like pattern and check whether I do any of it in my visioning actions.

Mirror Call: Authority/Citizen

His/Her BATHWAVE to Face and Embrace:
My self guidance scares me...
His/Her Quantum Shift to replace with their grace:
My self guidance can be confident.
My self guidance will be confident.
My self guidance is confident.

EVENING PRACTICE:
Tonight I embrace the mirror-calls of my AUTHORITY FIGURES, any of them such as teachers, bosses, church employees, government employees, etc.

First pick one that I have an issue with in the present. If none, then a past figure will do.
Talk to them in a mirror and tell them what I wish were different in relation to me. Write it down.

These positions are all mirrors of my self-guidance replacements. I would be their creativity.

Example: I wish my teacher was patient with me.

BATHWAVE OF THE MIRROR-CALL: Self-guidance (Authority Figures)

BATHWAVEs **Part 1: Face & Embrace**	**Quantum Replacements** **Part 2: Replace with Grace**
I face and embrace that my self-guidance is not patient with my creativity.	My self-guidance can, will and is not judging (patient with) my creativity.

I take a deep breath and feel free of judging my creativity!

Say my quantum replacements and abidingself sentence once out loud with my eyes closed.

DAY 7
AM I MY OWN BEST FRIEND?

MORNING PRACTICE: Say my quantum replacements and abidingself sentence once out loud with my eyes open.

Do I have any sisters or woman friends I can say something to that I wish were different in the mirror? Write it down. Cats also mirror your feminine.

Substitute my feminine side for her and whatever sex I am on my side. Hint: females are about opening sacred space, ignore this if I have my issue already.

Mirror Call: Woman with Cat

Her BATHWAVE to face and embrace:
I face and embrace that I am feeling victimized being a woman.

Her Quantum Shift to replace with her grace:
I can be safe as a woman.
I will be safe as a woman.
I am safe as a woman.

EVENING PRACTICE:
Every night I enter a dark room of my dreams where my eternalself speaks through my personalself. I have the ability through the dream to do and see who I am and what I am doing!
The dreams often use relationships to speak to me which is why I must learn this work first. Dreams give me the ability to change what might happen. They are not prophetic, yet give me options to transform myself.

Ask for a dream about a relationship I would like support in.
Say my replacements including new ones.

DAY 8
DO YOU RESPECT YOURSELF?

MORNING PRACTICE: Write my dream relationship down if I had one and replace the issue in the dream.
Say my quantum replacements and abidingself sentence once out loud with my eyes open.

Find a brother, or man friend. My masculinity is about my expanding active focusing. Embrace any issues with any male friends or brothers and replace them.

Mirror Call - Partners

Her BATHWAVE to face and embrace:
I face and embrace that I am irritated by my masculine beloved.

Her Quantum Shift to replace with her grace:
I can understand that my masculinity supports me.
I will understand that my masculinity supports me.
I understand that my masculinity supports me.

EVENING PRACTICE:
Once I understand dreams, I have the ability within a few weeks to change the dream in my daily life. I do this by doing something different, by choosing new constituents **instead of limiting myself to the constituents prior to the dream!**

List some new "things" I would like to learn or bring into my life and spend some time googling them to watch and enjoy.
Say my quantum replacements along with eternalself with my eyes closed.

DAY 9
DO I RESPECT ALL THAT I DO?

MORNING PRACTICE: Upon awakening, say my quantum replacements and abidingself sentence once out loud with my eyes open.

Now I can speak to a job, church, organization or culture I have issues with.

When I write the issue I will probably guess the aspect of myself that is being related to. Usually these will represent a nurturing or guiding figure again. If so, substitute with self-nurturance or self-guidance. Any aspect is possible so use my whole awareness.

His/her BATHWAVE to face and embrace:
I face and embrace that I am being really hard on myself...

His/her quantum shift to replace with their grace:
I can be gentle with myself.
I will be gentle with myself.
I am gentle with myself.

EVENING PRACTICE:
Ask for a dream of any relationships I missed replacing.

Say all my quantum replacements and code eyes closed.

DAY 10
DISEMBODYING RELATIONSHIPS

Now I will speak with "disembodied" aspects of me. Deceased family, pets or friends, angels or spirits fit this category. I replace the "disembodied" aspects with "embodied" aspects of me; such as, grounded, centered, or what I think would benefit my living!

MORNING PRACTICE: DISEMBODIED MEMORIES
After awakening, I say my quantum replacements and quantum code eyes open.

Now I will speak in a mirror to pets that have died that I miss, or spirits fit this category. I can speak to family and friends that have died that I speak to in my mind or dream of. I can speak to guardian angels, allies of native traditional cultures, anyone that does not have a body I can touch.

I am going to tell them what I wish were different in my relationship with them. I write it down. They are "disembodied" aspects of me.

I add the word 'disembodied' before whatever aspect of me that is spoken to in order to embrace and replace it with "embodied" words. Don't worry there are plenty of examples because this was a big issue in my living!

BATHWAVE #1

BATHWAVE OF THE MIRROR-CALL:	
BATHWAVEs **Part 1: Face & Embrace**	**Quantum Replacements** **Part 2: Replace with Grace**
I face and embrace the old habit of believing in disembodied concepts in my past and present.	I can, I will, I am fully embodied in this moment. There can, there will... there is no such thing as a dis-embodied particle/wave in the quantum world.

Even the quantum wave is embodied as a standing wave. It may be invisible, yet it is physical.

The quantum wave manifests the 'solid' matter world as I know it, which I think of as embodied. The wave spins when it turns around to go out again and the spinning wave appears to be a particle of matter.

Disembodied means I am talking to 'no quantum wave/no particle'. That's the recording of memory that I am talking to. I can only ask questions and read the answers. Next Decade teaches me more ways of doing just this.

Make a list of all people, images, and animals that I think about, or dream about, that are no longer living in a body I can touch today.

EVENING PRACTICE:
Do this as early in the eve as possible, as it takes time and energy.

Complete my list of all people, images, animals that I am thinking about that are no longer living in a body I can touch.

Just so that I know where I am going from a quantum point of view:

- **Everything that has ever existed is recorded in quantum memory.**
- **I have access to all of the past memory.**
- **I can ask questions and get answers that I may never have known before.**

In the past we have labeled people who do this as psychics, clairvoyants, and so on and so on. We all have this ability in quantum wave living and I will share with you later how to use muscle testing.

This informotion is not responsive like a person, though quite useful and supportive. How supportive it is, is up to me. It's about asking questions as I desire.

This recording is how I continue, though I am not continuous in my present body form.

When my personal body did not understand and live this, I noticed a detriment to my health and life.

I do not have to agree with this to continue. There is a lot of wiggle room in quantum wave living. Since I manifest it, I can have it any wave I desire.

If my particles disappear forever, it seems that so does my pulsating quantum wave. However, all my informotion of my patterns and my BATHWAVES remains recorded forever. All informotion is accessible by everyone from Universal Studio's quantum recordings, free for the asking.

This is how I continue in this moment of awareness. This has been verified by using muscle testing and other methods.

I will receive many examples and then continue with the tools I have learned.

If I do not wish to do it now, then put a bookmark here and return when I complete the book or better yet, find a quantum living coach to guide me, even online!

When I have tears from each eye, I remember to put them under my tongue as a homeopathic remedy for what I am feeling!

Examples
I look in the mirror and speak to my grandmother who was my first acupuncture client... "I am missing YOUR trust and unconditional love, grandma!"
Grandma is a mirror of my **WISE SELF-NURTURANCE**. I am her grandchild, so **grand creativity**.

BATHWAVE OF THE MIRROR-CALL: Wise Self-nurturance & Grand Creativity	
BATHWAVEs **Part 1: Face & Embrace**	**Quantum Replacements** **Part 2: Replace with Grace**
I face and embrace that my grand creativity misses the trust and unconditional love of my wise self-nurturance.	My wise self-nurturance can, will and does trust and love my grand creativity unconditionally in this moment.

I feel more centered and embodied with just this one.
Do any personal BATHWAVEs as they come up. Write the replacements down in your notebook.

Religious example of unmet needs:
Continue with the list I made when growing up. I spoke to my guardian angel and the image of Mary, mother of Jesus, a large portion of my childhood.

BATHWAVE OF THE MIRROR-CALL: Religious unmet needs	
BATHWAVEs Part 1: Face & Embrace	**Quantum Replacements Part 2: Replace with Grace**
I face and embrace the unmet need for a caring, listening, unconditionally loving, nurturing self for my creativity.	I can I will and I do care, listen, unconditionally love and attend to my creativity.

My deceased mother:

BATHWAVE OF THE MIRROR-CALL: Deceased Mother (Self-nurturance)	
BATHWAVEs Part 1: Face & Embrace	**Quantum Replacements Part 2: Replace with Grace**
I am depending on touches and hugs of self-NURTURING in order to like myself.	I can I will and I do like and feel warmth with my being, hugging/nurturing my self.

My deceased father:

BATHWAVE OF THE MIRROR-CALL: Deceased Father (Self-guidance)	
BATHWAVEs Part 1: Face & Embrace	**Quantum Replacements Part 2: Replace with Grace**
I face and embrace that I am dependent on the admiration and acknowledgement of my self-guidance.	I can I will and I am embodying the acknowledgement and admiration of my self-guidance.

Record and do any BATHWAVEs as they come up.

EVENING READING: TELL-A-VISION DREAMING NEXT Decade

One way I do my visioning is in my dreaming and in this I begin to exercise and understand the communication between my quantum wave-like pattern and my personalself.

This is what the dream is; my abidingself communicates through my personalself to give me options. This interaction is the manifestation of my resplendent illuminatedself.

The dream is an ingredient, in and of itself, for one of my interactions as my resplendent illuminatedself.

My personalself interacts and chooses new ingredients for what nourishes me in my wisdom creating an emergent story to be new and different and always radiant!

TASTE FROM WHAT'S NEXT...QUANTUM DREAM Decade

I had a dream the night after WRITING this revealing another BATHWAVE RELATING TO DAD. "A papa jumps into a well. His daughter and I are watching. His hands reach up to waters edge and we reach in and pull him out."

BATHWAVE OF THE MIRROR-CALL: Deceased Father (Self-guidance)	
BATHWAVEs **Part 1: Face & Embrace**	**Quantum Replacements** **Part 2: Replace with Grace**
I face and embrace that I am holding on to emotions around losing my self-guidance. I feel sadness, fear and blame.	I can, I will, I am safe, secure, and confident with my self-guidance, blessing those around me.

I have always had emotions around my father's death and I blame others.

Congratulations on completion of understanding basic relationship mirror-calls (miracles).
Onward to quantum dreamwaving!
Before I sleep say my quantum replacements and abidingself sentence once out loud with my eyes closed.

QUANTUM REPLACEMENTS FOR RELATIONSHIFTING

Please say your relationship replacements from your notebook.
These below are optional from the examples. I will repeat these present tense replacements for at least 21 days to 3 months.

My self-nurturance appreciates my creativity as it is.

My self-guidance connects and communicates supporting my creativity.

My self-love is new and different each moment appreciating and cherishing me passionately.

My male creativity is active, expressive, expanding and focusing.

My female creativity opens sacred space.

My self-guidance is not judging (is patient with) my creativity.

I am fully embodied in this moment.

There is no such thing as a dis-embodied particle/wave in the quantum world.

My wise self-nurturance trusts and loves my grand creativity unconditionally.

I care, listen, unconditionally love and attend to my creativity.

I like and feel complete with my own being which nurtures myself.

I am embodying the acknowledgement and admiration of my self-guidance.

I am safe, secure, and confident with my self-guidance, blessing those around me.

Introduction to dreamwaving with Kathy's Dream:

"Ruth (a friend) and three young boys were in a hardware store. Ruth was opening a set of screwdrivers with the boys. I saw Ruth. I walked behind the aisle they were at, observing from a distance. I too needed a screwdriver, yet I did not want to interact or be seen by Ruth and the three boys.

Kathy's Perspective Dreamwaving:

My feminine and creative selves have issues engaging and interacting with my self and friends.

Kathy's Quantum Replacement: I can, I will and I am opening my femininity and creative selves to engage and interact with my self and others."

Dr. Angela's Perspective Dreamwaving:

Screwdriver = fixing things

You may feel that you have to fix others. People do not need fixing; you cannot fix people.

"After this session with Dr. Angela, I felt some discomfort when I returned home. I continued to identify areas in my life where I felt the need to fix either myself or other people. I worked for quite some time; identifying the frozen age of 10 years old. I had wanted to fix peers of mine and I am realizing that the failure of not being able to "fix" them was also an old pattern I had been carrying around. Carrying around this belief, this understanding that I was a failure was huge. Here I had unknowingly been carrying around this heavy weight of failure for 25 years. Each year this pattern was growing heavier and repeating over and over.

This understanding of the fixing pattern in my life made a huge shift for me. It was the first time I was aware of how much I try to fix people and that failure was crippling me. This awareness was powerful, realizing that people including myself do not need to be fixed and cannot be fixed. We are not broken."

DECADE 5
QUANTUM DREAMWAVING LOVE LETTERS

Dreams are a direct hyper-communication from my eternal abidingself to my personal being - a veritable love letter! It tells me what I am doing in this moment or the next couple weeks so that I can choose new options when I desire.
Yes, it is there to assist my living!

Quantum Tip!

The first image / action that I recall is a holographic key to understand the whole dream

Everything in the dream is an aspect of myself

DREAMS AS LOVE LETTERS HAIKU

Dreams are scenes of next
Two weeks' doings without
Judging! Love new options…

DAY 1
DREAMWAVING TECHNIQUE

AUDIO RECORD YOUR DREAM and/or WRITE IT DOWN

I only need one image PLUS one action from the dream, the last one or the first one is preferable, however, any will do.

Upon awakening, RECORD my dream immediately! I describe all that I remember with every detail like colors, people, ages, and setting. Everything I experience is relevant such as feelings or thoughts too. Audio is best as writing may compromise the images.

WHEN I DO NOT REMEMBER MY DREAMS, no worries. I can use an event from the day before that I remember and would like to understand or transform. Just write down the image of the event and the first full action. Then treat it just like a dream with the process laid out from here on. After steps 1 & 2 if I run out of time I can continue in the evening.

MORNING PRACTICE: Say my quantum replacements and abidingself sentence once out loud with my eyes open.

1. Write down the first or last image and action of my dream that I remember. It does not matter if I forget any parts of my dream. If I have no dream recall use an event from my day.

____

____

____

____

1B. It's all about me. Everyone in the dream or event is an aspect of me. Notice, muscle-test or use GAME to identify what aspect of me is represented in the dream by the images or people that appear in it (See Fig 4.1 for relationship mirror-calls). List those dream images with their meaning. Use the Table 5.1 for guidance.

____

____

____

____

2. Rewrite #1 with replacement meanings. Choose a meaning or message about my action. Muscle-test, use GAME or notice if my meaning makes sense to me.

____

____

____

____

Table 5.1

Quantum Tip!

General meanings of common images in dreams:

Image:	**Represents:**
Water/fluids:	Emotions
Tree:	Their material wood changes to "would"
House and room:	Your 'self'
Bed:	Relationship
People:	Mirror-call aspect of you Table 4
Car:	Physical body
Age:	What happened when you were that age
Animal:	Meaning you give them in waking state
Dog:	Friendship; unconditional love
Cat:	Femininity
Birds:	Ideals; above it all, purpose

3. Is this the meaning I desire? Is this the doing I want to continue in my life?

____

____

____

____

4. If my answer to the question 3 is No then face and embrace my action and replace it with the desired one. Write down my replacement or comments.

____

____

____

____

EVENING PRACTICE:

Finish my dream or daily event understanding and replacements. Add quantum replacements to my existing list. Continue looking at this dream below from a broad perspective of my living.

5. Notice, use GAME or muscle-test the new option after I replace the old one with midline quantum movement. Is there more I would like to add? Write the replacements down. Notice/muscle-test/use GAME to find out if the replacements need maturing (See Melting Frozen Ages Method Chapter). Write down my quantum replacements.

___

___

___

___

6. Review the dream from a quantum transformed perspective. Does the rest of my dream makes sense now? Do I know why I gave myself this dream?

___

___

___

___

Say my new list of replacements out loud before sleep along with the old list.
Ask for a dream about a topic I would like. Be ready to record it with voice memo or write without much activity or light. Have phone or paper and pen by my side.

DAY 2
DREAM EXAMPLE

MORNING PRACTICE: Record my dreams first. I can do the process from yesterday or read the example. Say my quantum replacements and abidingself sentence once out loud with my eyes open.
Example of a clients dream:

1) *Write down the first image and action of my dream that I remember or the last. It does not matter if I forget any parts of my dream.*

 First image and action:
 "I am in a Chalet on top of a mountain. I am moving my suitcases around in the reception area."

1B) *Remember it is all about <u>me</u>. Everyone in the dream or event is an aspect of me. List those dream images with their meaning. Use the Table 5.1 for guidance.*

The **Chalet (Hotel) as a building** would be an image of her **"self"** which many other people are staying in.
I asked what the **mountain might portray for her**? she answered **"my higher self..."**
Baggage = what she is carrying around in her living, (slang use of word baggage).

2) *Rewrite #1. with replacement meanings. Choose a meaning or message about my action.*

The meaning after substituting all the words above:
The Chalet: Since many people other than herself stays in the Chalet, her dream might be saying that her idea of her "higher self" is composed of other people's philosophies? Not her own self.
Moving her baggage: All she is doing in this situation is moving her baggage. **She is not sure of wanting to move into this old self?**

3) *Is this the meaning I desire? Is this the doing I want to continue in my life?*

She laughed and said this felt accurate.
I asked her if she wanted to continue this way? She said no.

4) *If my answer to the question 3 is No. Then I face and embrace my action and replace it with the desired one.*

She embraced this and replaced it with: "My quantum wave pattern abidingself is freeing and guiding my living as my emergent unique illuminatedbeing."

5) *Notice or muscle-test the new option after I replace the old one with midline quantum movement. Is there more I would like to add?*
Notice/muscle-test/use GAME to find out if the replacements need maturing. Write down my quantum replacements.

No maturing is necessary.

6) *Review the dream from a quantum transformed perspective. Does the rest of my dream makes sense now? Do I know why I gave myself this dream?*

In hindsight, she had informed me that she had stopped all practices she had been doing for 15 years feeling stuck in the old way that was not satisfying her need for expansion. So the dream made sense to her.

Her unique and individual active quantum wave pattern called her eternal abidingself is a guide with personalself expanding everything she does. Guiding her with her dreams, relationships, symptoms, events, creations, everything. What resplendent living!

Remember that dreams are showing me what I am doing, for my benefit, even nightmares. In the case of nightmares, my living may look like a horror show. When no options pop up, I can find options I desire to replace it with using the quantum mirroring technique from the last Decades in this moment

EVENING PRACTICE: MY TRANSFORMATION Haiku

My transformation
Is my invitation to have
The world my wave!

First finish the work with my dream from this morning.

> "...for, in the actual universe, both me and my journey are emergent, new and different replacing the previous me, as I go along!"
>
> "Practice accurately visually remembering, and recalling dream images! ...every night, I have four or five dreams of my eternalself, in the actual universe, both you and the journey are emergent or new and different in each moment as you go along!
>
> Practicing as we have suggested is more than training my mind ...or sensing and feeling, it is to be very real and literal ...or it isn't practice at all! ..."fake it 'til you make it' will not work for you!"
>
> In introducing you to these practices, we are inviting you to expand and extend my personal self and life today by calling forth my eternalself to communion with my personalself and life, ...so you can live and portray what has not been seen for a very long time! ...how would you like to portray your eternalself?" Cotting, R. B., (2012), Pericope 14a #43

Ask for a dream image of my quantum wave pattern and whenever I wake up record it before turning a light on no matter how little I remember.

Say my quantum replacements and abidingself sentence once out loud with my eyes closed.

DAY 3
DREAMS SPEAK THE QUANTUM LANGUAGE OF VISIONS

MORNING PRACTICE: Record any dreams you remember and do as much of first day process as possible.

Say my quantum replacements and abidingself sentence once out loud with my eyes open.

> "Pay close attention to your dreams! ...for your dreams are speaking to you in the language of visions! ...most of the words you hear or sense hearing in a dream are

your own comments ...and are not part of the dream! ...you certainly want to see what you as an eternalbeing are really doing and becoming instead of relying on your own judgments and commentaries ...isn't that right?!" Cotting, R. B., (2012), Pericopel4a #7.

EVENING PRACTICE:
First finish the work with my dream from this morning. Then reread yesterday's evening practice.

Say my quantum replacements and abidingself sentence once out loud with my eyes closed.

DAY 4
NOT HAVING FEELINGS RUN MY LIVING

MORNING PRACTICE: Record my dream or an event you would like to understand and quantum process with it. Then say my quantum replacements and abidingself sentence once out loud with my eyes open.

"Flow my thoughts as images! ...when my mind is filled with thoughts, do not stop them or slow them ...expand them until you are breathless! ...let them glow as a pathway of images leading you directly to my eternalself and the inspiring self and life you truly desire!" Cotting, R. B., (2012), P. 14a.

Today I practice thinking in images and describing these images whenever I can.

EVENING PRACTICE:

"Practice living every moment by promises and dedication! ...you are addicted to living by feelings, feeling my feelings, and reassuring myself by my feelings! ...unfortunately, being spontaneous instead of being myself confines you to judging and reacting ...which is slowly destroying you!

Your entire life as an eternal abidingbeing is built upon promises ...while my life as a personalbeing seems to be founded on impulses! ...when you as a personalbeing forsake my promises and give in to impulse to be spontaneous ...and do what you feel like doing instead of keeping my promises! ...this tends to induce underlying feelings of guilt and remorse ...and a bit of shame and discontent! ...and you feel uncomfortable, such that, to feel better, you do even more things to feel good about myself again! ...but beneath these feelings is an awareness of promises and dedications you are neglecting! ...and as you continue this conflict between feelings and promises, my immediate feelings almost always triumph over long-term promises!

What eternal promises are you struggling to ignore or deny or cover with personal feelings? could it be that the fear of uncovering these promises keeps you from truly being aware of my abidingself and knowing myself as a communioned

> personal-abidingbeing? how long will you continue this drama or charade? or have you come to depend on this conflict to add excitement to my life?" Cotting, R. B., (2012), Pericope 14a
>
> In Roger B. Cotting's writing he speaks to two of my triunity with pronouns: roughly speaking I believe 'You/your' refers to my personalself and 'me/my' refers to my abiding/quantum wave pattern self. Don't worry. It's not a big deal.

I write QUESTIONS or THOUGHTS about this radical new reading. If I see any patterns I would like to replace about THESE IDEAS or about feelings, I do that. I write my replacements in my notebook Feel free to disagree.

It is a relief to not have feelings running my living!

Say my quantum replacements and abidingself sentence once out loud with my eyes closed.

DAY 5
QUANTUMCODE IMAGE OF ME

MORNING PRACTICE: Repeat the process with my new dream or a different daily event. Follow the steps from yesterday. Write my dream down in my dream journal. Or record it verbally on my voice memo on phone.

Say my quantum replacements and abidingself sentence once out loud with my eyes open.

Do I see how my dreams are really love letters to myselves? They are set up like a lover to show me what I am doing to give me new options!

Only a lover would go through all of this for me!.

EVENING PRACTICE:
Please reread yesterday's eve practice. Ask for a dream about a topic I would like assistance with. Before I sleep say my quantum replacements and abidingself sentence once out loud with my eyes closed.

DAY 6
CONVERT EVERYTHING I HEAR TO IMAGES

MORNING PRACTICE: Say my quantum replacements and abidingself sentence once out loud with my eyes open.

> "Listen with my eyes! ...forget all about having the ears to hear ...or listening behind the words! ...what is important is whether or not you convert everything you hear

to images, like dreams, and follow those visions to picture and understand what is being said!" Cotting, R. B., (2012) Pericope 14a

Create pictures of all my communications today. Jot them down to remember them in order to describe them like a story, in a writing I will compose this evening.

EVENING PRACTICE: **IN MY DREAM HAIKU**

In my dream I am
Aware and see options as
Resplendentbeing!

Write the story of my day's communication in pictures. It does not have to make sense. It will make sense as a dream does. A dream consists of a series of images necessarily related though not obviously connected as I use brain thinking.

Take the story and treat it as a dream. Use the first part of my initial communication as my first image and first action.

Replace whatever perception I wish. Write down my new replacement. If there are none congratulations! Sweet dreams! First say my quantum replacements and abidingself sentence once out loud with my eyes closed.

DAY 7
NOTICING DOING DREAM OPTIONS WITH FEW WORDS

MORNING PRACTICE: Say my quantum replacements and abidingself sentence once out loud with my eyes open.

Here is another sample dream by a student of quantum living to give me more experience.

IN TOM'S RESPLENDENCE, A DREAM TRANSFORMS A DAY...A LIFE?

"I had a dream, my best friend and I were in a car, and it had water in it, inches deep inside. So we had to tip it to get the water out.""

I was thinking how much my friend looked like my father these days. My father is a big man, physically strong, with presence: quick to laugh, caring, and quick to judge, too.

In my waking day I practiced growing and gathering desirable thoughts and feelings, savoring moments of beauty, accompaniment, small things all that enriched and filled me.

I changed my routine, and took my lunch in a park, in the shade of a tree on a hot sunny day. After I finished my lunch, I lay back on the grass and enjoyed the feeling of relaxation and support from the earth.

I felt buoyant, refreshed and navigated through the city traffic to a parking spot right outside my meeting, such a rare blessing in downtown Sydney!

After work I took a little scenic trip on my way home, along the beautiful waterways near where I live. Soaking up the beauty of the mangroves, the light on the water, the pelicans...I went past the Sea Scout Hall, and there were two long canoes each with about eight little Joey Scouts around six years old. They were paddling like mad to try and get going off the sand into the headwind. They were excitedly laughing and seemed full of light-hearted determination. In the face of no visible progress they kept on going and were loving it.

I cycled home, happy, touched by that, and all the richness of not just another day."
Christiansen, T.

I spoke with Tom that evening about the dream and mentioned that the car was his physical body. The water is an emotion, which was determined to be sadness. Since he said the friend reminded him of his father or could be his fathering in some way. During his day he was befriending that 'sadness' and replacing it with new attitudes and actions that were transforming his 'sadness' to emergent miracles he enjoyed!

He had resplendently understood the dream's message.
He realized that he could not pour out the emotions from his body as he tried in the dream. He replaced them with new attitudes, actions and emotions!!!

His daily event at the ocean with the wind (which represents judging) stopping the spatial progress of the paddling children, might be a little mirror-call that he still has been a little sad about feeling judged by his dad as a child. Of course from *a quantum perspective that would really be that his self-guidance judges his creativity, with the belief that he is not going anywhere in his life.* **He already replaced that with his response to this event with "lighthearted determination" and "just loving it!"**

Bravo Tom!

I do something different today that I desire, however small.

EVENING PRACTICE:

> 20: "my dreams are one continuous saga of myself and life ...they are not a feedback process for you today... In fact my dreams are a true picture of my self and life!... Do you understand the real meaning and significance of your dreams?... And the truth of death being portrayed?... You are always being prepared!"
> 21: .... You (I) do not control my body!... Or even direct my body!... You direct the self that directs the body you have!... This is selfing!"

99 Matters That Matter by Cotting, R. B., 2010.

The self is like my quantum wave and the body is like the particle. **Remember that the quantum wave makes a new particle at its center every pulsation.**

Pulsations are as fast as the speed of light?! New informotion transforms the particle instantly!

I direct my quantum wave as I am living the informotion I desire. Living new and different informotion gives me a new and different body. How I am living is also how I am dying. My dreams reveal how I am living, therefore prepares me for dying!

Ask about what I think my death will be like and tomorrow write about it based on my dreams or if no memory of dreams, use any BATHWAVEs I am living with in the morning.

Say my quantum replacements and abidingself sentence once out loud with my eyes closed.

DAY 8
MY LIVING IS A PICTURE OF MY DYING

MORNING PRACTICE: Say my quantum replacements and abidingself sentence once out loud with my eyes open.

Write about what I know about my dying from my dreams or my daily living. Then process my dream or a daily event from yesterday, quantum style.

EVENING PRACTICE: My evening practice is to look back at my day today. Write about any changes I have made today reflecting on my dream or daily event. Did they mirror the meaning of the dream or event at all?
Mention any other BATHWAVEs I might like to replace from this exercise.

Ask for a dream tonight that will continue assisting in transfiguring some part of my life I desire! Say my quantum replacements and abidingself sentence once out loud with my eyes closed.

DAY 9
SENOI INDIAN DREAM CULTURE

MORNING PRACTICE: Say my quantum replacements and abidingself sentence once out loud with my eyes open.

After I understand my dream's love letter to myself, I am going to read about a contemporary tribe in Malaysia. The tribe is called Temiar Senoi whose life is focused on their dreams. Their tribe has minimal neurosis and conflict even though they are surrounded by fighting tribes.

The reason I chose them is that there is no separation between dream and life. The whole community is involved. The tribe is handled with images and actions and creative products from the gifters in the dream to expand awareness and pleasure.

A summary of the Senoi Indian dream living system is:

1) Advance toward **dream pleasurable experiences** extracting a **creative gift** to be **shared** when a **person** is **involved**..
2) When you **fall,** have **no fear and explore** where you are to **share** in waking time.
3) **Confront** and **conquer danger** and when an **opponent** dies they become a **friend** who **gives** you **a gift** to **share.**
4) Make my **dream** a **positive** outcome and extract a creative product from it.
5) When there is **difficulty** in **relationship** with a **person from my community in my dream, inform** them and **act** on it in **waking time...**with **gifts** going either way. Garfield, P. (1981)

These guides can be integrated into this quantum system after the initial process is completed.

Begin sharing a dream with at least one friend. Doubly better if they are also reading this book. I am willing to listen to their dream too. Relate to the dreams with a quantum and Senoi perspective. If no friends yet, just write and share in my journal as my resplendent illuminatedbeing, witnessing my interaction of my eternalbeing and my personalbeing.

EVENING PRACTICE:
Read over the Senoi Dream living system again and ask for a dream I would like, based on their system.

Say my quantum replacements and abidingself sentence once out loud with my eyes closed.

DAY 10
SUMMARY OF DREAMWAVING: MAGINATION IS VISIONING

MORNING PRACTICE: SUMMARY OF DREAMWAVING
Say my quantum replacements and abidingself sentence once out loud with my eyes open.

Dreams are **quantum thinking** with the universe about **what I am doing** in the moment. Then I have the ability to **replace with** our **true desire** options. Dreams are our **sacred guide**.

The quantum method of understanding my dream communication helps me prevent the judging of languaging. I look at one image with one action as an aspect of my selves without judging. If I want something different than what the dream reveals that I am doing, I can do another option and have a different emergent in my daily living. It's as simple as pie. There is no judging by the dream, so I will not add judging to my dreams!
The quantum way of "thinking" with the brain of the universe is NEXT!!

EVENING PRACTICE: CONVERSATION ALL ABOUT ENTANGLEMENT*

Roger: Having this understanding (OF ENTANGLEMENT) gives you some thing to work with in life. Without this, everything that's conjured up is confusing. Space used to be defined by science as nothing.

(Angela adds from research reported by Professor David Tong from Cambridge, "Now we know from the quantum world that space is full of "quantum fluctuations". See YouTube "Quantum Fields")
Dr. Angela Longo believes these quantum fluctuations are related to entanglement.

Angela: Doesn't it mean that memory exists within space?

Roger: Space **is memory** itself.
It's the memory between things that is gravity.

And it's the relationship between things that expresses time or space time.

Angela: So, is entanglement memory?

Roger: Yes.

Angela: Wow. Entanglement is memory!!! (Then **space is entanglement!**) This word 'memory' does not mean brain memories.

Roger: It's in book one, chronicle A. That's discussing Einstein's Spooky Action at a Distance theory and what was going on was memory.

Angela: Is 'entanglement' how we access the information of the universe?

Roger: Yes, you are accessing memory and that's why I call it 'memory'.

Angela: It has a quality like memory accessing informotion.
That makes sense then. (This is the quantum memory of the universe's informotion.)

Can we direct entanglement?
Can we ask for a particular entanglement?

Roger: Can you ask for a particular memory? Yes.

That's called direct perception and that's what maginations are all about.

You are accessing it directly. That's why it says to have a focus.

Without a focus, you wouldn't know where to go or you wouldn't understand a thing.

So the focus in magination is directing the memory that's part of the magination pages*.

Angela: So the focus of magination directs entanglement and that is how magination works. Amazing!?

Roger: Direct perception has a focus, and it's the focus that directs the perception. It enables you to understand what you perceive. **If you didn't have a focus, you wouldn't understand what you perceive.**

Angela: That makes sense.

Roger: Because you perceive a whole, which wouldn't make any sense unless you had a focus to understand the whole.

Angela: When you say a whole, what do you mean?

Roger: It's a whole event. It's like information. You only understand information because you get it a little bit at a time and you see it unfold. If you saw all information unfolded at the same time, it wouldn't make any sense to you. It makes sense because you unfold it according to language. You decode it either way according to language, which of course, screws it up.

Angela: So, it's like a dream. Better not put words on it.

Roger: So direct perceptions are like dreams.

Angela: Don't dreams speak in whole events?

Roger: Yes, yet you only understand dreams pieces at a time as the dream unfolds …because you're accustomed to dealing with memories.

Many of these ideas Roger proposed in his resplendent magination way have been demonstrated by scientists since then.
*Transcribed conversation in 2013 between now deceased Roger B. Cotting (1931-2017) and Dr. Angela Longo

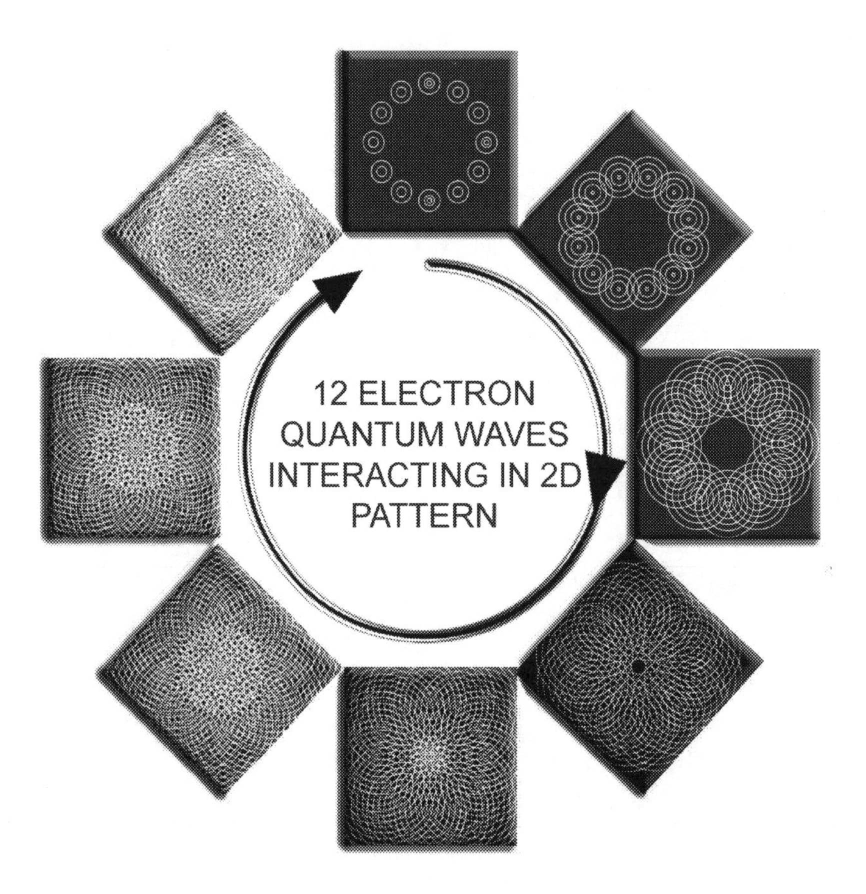

Years later, Angela's afterthoughts:
Entanglement sculpts energy though it does not require energy. It is almost like a free moving kaleidoscope. It is transfiguring interacting quantum wave (pulsating) images without adding or subtracting anything! Since space is filled with these interacting quantum waves, imagine what this space must look like then!!?

Maybe like a kaleidoscopic show of 'colorful' quantum wave informotion interactions like mandalas? This somehow contains the informotion of all the happenings of my self and the universe and I have free access to it! I call it **GOOGLEVERSE!**

So every intelligent being in the universe is in entanglement with me and vice versa. We all benefit from each other and I can unentangle the entanglement when I use magination. I am focusing on my abidingself with my visionary ability.

Space or entanglement is the effortless connection or interaction between everything, that I have mistakenly called "oneness", which is the farthest thing from the truth. It is the very distinct individual uniqueness of each of our particle's quantum waves that empowers entanglement! I can say I am three selfs in one-nest.

I let these images and ideas play in my imagination. How might they relate to me? Whatever I remember from this reading I write about impacting my living!

Review morning reading and write notes about how the reading on entanglement might relate to myself today.
Say my quantum replacements and abidingself sentence once out loud with my eyes closed.

"Can I ask for a particular memory? **Yes.** That's called direct perception and that's what maginations are all about. I am accessing it directly. That's why it says to have a focus." Cotting, R. B., (2012)

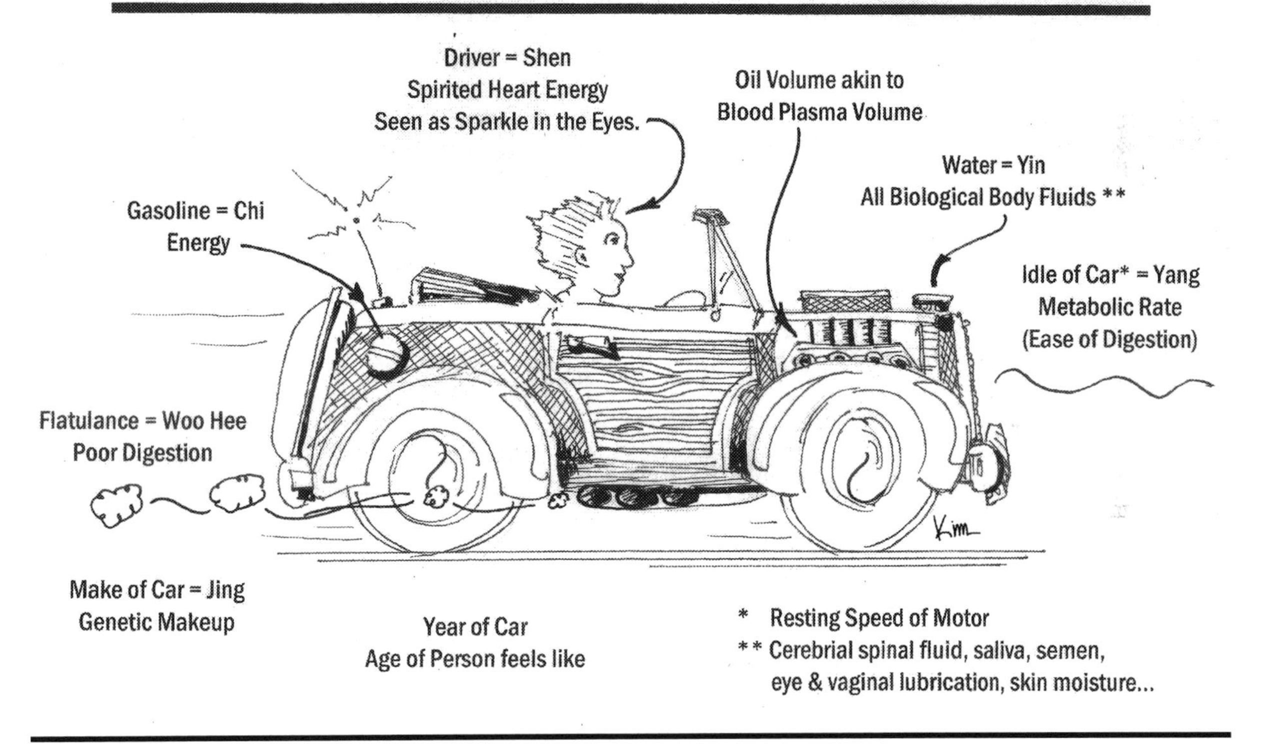

DECADE 6

GETTING TO KNOW ME. GETTING TO KNOW ALL ABOUT MY BODY/ENERGY

DAY 1
I AUTO KNOW

MORNING PRACTICE: Say my quantum replacements and abidingself sentence once out loud with my eyes open.

Not knowing about your energy is like living in a house without knowing that it has an electrical system. An example might be that I am not aware of burnt out light bulbs, fuse boxes and possible short circuits in my house. And that is how I was living most of my life and my health suffered. I ignored all of the signs till I would be almost dying.

Likewise, would you never give your car a tuneup? It seems to me that the ancient Chinese were probably aware of this 'wholographic' universe in using these five elements' relationships in medicine. Before I begin this explanation of the elemental relationships, I would like to mention the five 'substances' diagramed in the lovely auto on the opposite page.

The metaphor describes the Blood (yin and yang aspects), the Chi (energy), the Yin, the Yang, and the Shen (the sparkle in the eyes of the driver). Giving my car a tune up would be like acupuncture or an acupressure massage. Putting in oil/water and gasoline would be like taking herbs and appropriate nutrition. Relationshifting would be giving attention to the driver. Isn't the driver in charge of taking care of their car? If only I took as good of care of my body/energy as I do my car, I would be in great shape.

Be aware of being in charge of my body and all decisions around my body. Am I able to do this throughout my whole day? What is my issue if I cannot? Take notes and replace it when possible. If my car has issues what part of my body does that relate to looking at the auto diagram? Shift that now or later.

EVENING PRACTICE:
Study the car diagram "I Auto Know…" and attempt to learn the car aspects of my body. Ask for a dream about my own car, even if I don't have one, to help me understand my body. My biggest nightmare as a youth was driving a car and losing control of the steering threatening a collision. If only I understood that my eternalself was telling me that I was not driving, therefore, not in charge of my physical body.

Say my quantum replacements and abidingself sentence once out loud with my eyes closed.

DAY 2
WISDOM OF THE BODY

MORNING PRACTICE: After any dreamwaving, I say my quantum replacements and abidingself sentence once out loud with my eyes open.

From my BATHWAVES I ***outpicture*** my life, my body, my health, my well being or my illnesses, and disorders and disabilities. It is literally the informotion of the BATHWAVES. My physical body and wellbeing are being patterned by me in the moment and they're a confirmation of all I am BATHWAVEing about my self and others. My body is an image, a mirror-call image of me as my way of life. My body's well being or seeming frailties have nothing whatsoever to do with luck or misfortune. They are an instant outpicturing of my patterns.

One client felt a sudden pain in her front chest when we were working together, which instantly disappeared upon replacing the relevant BATHWAVEs. That made a believer out of her.

The universe supports and accommodates what I desire, what I am living and offering. Free will is pre-eminent. My body is a living picture of what I am doing. Shift what I am doing and I will see changes in my body immediately, not over time like in cause in effect. However, old chronic conditions may be layered and take time. Health is only a byproduct of this wave living, not the goal.

Study the MRI chart and see how the meanings make sense of the various body parts. Throughout my day be aware of the outpicturings that occur. Keep note of the outpicturings to work on later. The MRI chart that I see in the figure shows me words that will be useful in translating the love letters of aches, pains and diseases.

It would be good to begin learning the left and right and other body parts that make sense to you.

EVENING PRACTICE:
In any relationship even with our body, it is not the condition of the body that is the mirror, the mirror-call is my response or reaction to what is going on that is about me.

Using Relationshifting is quite an exciting form of alternative medicine that is called Epigenetic Medicine because it's outside of the "genes"; even though the "genes" may be the vehicle for some of the entanglement. It is not the central dogma of biochemistry anymore; I used to think of it as an equality of gene to body part and/or condition. I no longer I believe that and many of my colleagues agree. (Lipton, B., 2005)

With these new understandings it becomes important to begin simplifying and forming an image of the lifestyle I desire and live that. Living as I desire is what quantum wave pattern is all about. To do this I need to move my focus back to my abidingself as I am able.

Say my quantum replacements and abidingself sentence once out loud with my eyes closed. Ask for a dream that will assist me in focusing on my abidingself.

HEAD ~ Consciousness, embodied thoughts

NECK ~ Transforming and embodying awarenesses and mindful thoughts or that which connects my thoughts and dreams to the body of my life. Pains in the neck are disbeliefs that I can live our dream and my thoughts.

SHOULDERS ~ Carrying commitment choices, duties, responsibilities, burden

ARMS ~ Reaching out, putting choices into action, giving to life

HANDS ~ Holding, handling, offering, receiving, acting upon

ELBOWS ~ Flexibility, uniting powerful and exciting actions ~ Joints are flexibilities in my lives. Swelling would mean emotion blocking the flexibility.

FLUIDS ~ The emotions of my life

BIOLOGICAL FLUIDS ~ like blood, Spinal fluid, saliva, tears, etc. ~ Vital emotions

FLUIDS OF SICKNESS, LIKE MUCUS ~ Undesirable emotions that don't serve us anymore.

CELLS ~ Beliefs.

BONES ~ Convictions.

LUNGS ~ Attitude

INHALING ~ Taking in, vitalizing and inspiring attitude.

EXHALING ~ Releasing undesirable, inappropriate attitudes, judgments or other matters.

HEART ~ Empowerment of emotions and feelings.

ABDOMEN/DIGESTION ~ Examining and assimilating nourishing matters.

HIPS ~ Putting self-empowerment into action.

LEGS ~ Motives, desires, intention, purposes.

KNEES ~ Flexibility, uniting our intentions and purposes.

ANKLES ~ Flexibility, uniting, understandings, desires and intentions

FEET ~ Understandings.

TOES ~ Reaching forward with my understanding.

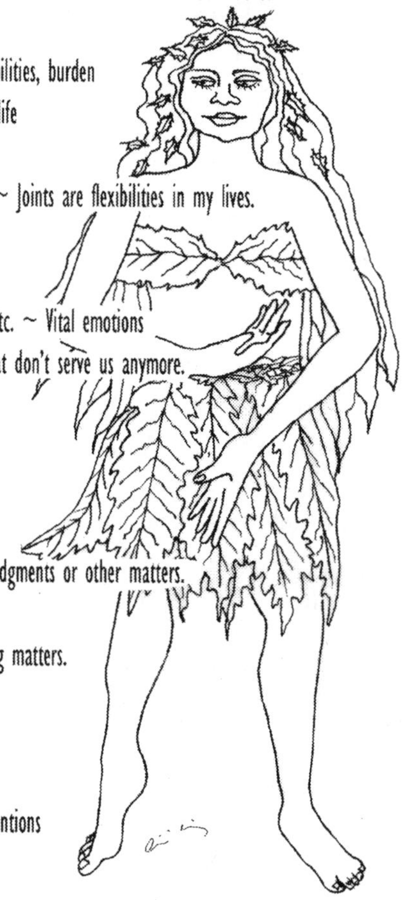

BODY ~ mindful, spirited, fully embodied.

BACK ~ the past.

SIDES ~ the present.

FRONT ~ the future.

RIGHT SIDE of the body ~ appropriate, worthy, good, desirable or concept of right.

LEFT SIDE of body ~ inappropriateness, wrongdoing, unworthy, undesirable, bad, and no.

MRI=Mirror-call Reflecting Image

DAY 3
WHAT THE PAIN...MRI

MORNING PRACTICE: Say my QC & R.

The descriptionary is concerned with the BATHWAVEs (beliefs, attitudes, thoughts, habits, words, actions, values and emotions) I'd choose as a way to activate and express myself in life.

My outpicturing of beliefs inpictures myself. The inpicturing of myself outpictures my beliefs. Einstein knew that the observer influenced all they were observing.

My outpicturing of my body is what I am BATHWAVEing, which is beyond even the concept of influence. It is the entanglement and emergent nature of reality. My outpicturing, whether of illness or well-being, simultaneously outpictures the inpicturing of what I believe, think, feel, say and do.

It's the actions themselves that are important, not the what's or the why's– although some of the why's may give us clues as to the how's of the actions.

This may seem involved and complicated. Simply reading the love letters that I give myself in my body really are gifts of my resplendence.

One of my clients, Carla, in New Zealand came in with a rash. Her rash had been around for two weeks and was very disturbing, red, welty and itchy. By the end of our work session, she came up with the correlation the how's and why's of her actions and "rashionalizing" in her life, and clearly shifted everything. She walked out 90% better and by the next day the rash was gone. Some of her BATHWAVEs were quite straightforward and easy, since she could pinpoint when the rash had started. Using common sense will serve well in seeing what BATHWAVEs are needing to be shifted.

Observe myself throughout my day with the question of what percent of my day is being lived as my abidingbeing.

EVENING PRACTICE:
Mirror-calls offer the keys to prevention. When I catch conditions in the beginning, which is what TCM, Traditional Chinese Medicine, always attempts to do, I can prevent harsh diseases from occurring. For example, my liver cancer diagnosis was a lot of the repression of anger that I had done in the first 50 years of my life. Anger is the main emotion of the liver. Through years of BATHWAVESs, herbs, and acupuncture, but BATHWAVESs mostly, and Relationshifting of course, the liver cancer diagnosis was replaced.

Do I more fully understand how–my outpicturing–consisting of my BATHWAVEs as in Carla's case, are my symptoms? And that I am rashing those same symptoms in every aspect of my self in life? Take heed of this awareness and stop the BATHWAVE of rashing. When I treat a physical rash while I continue the activities of rashing, I may soon have a severe rash on my face, which indicates that I am irritated by having to face my irritation. Or, the face rash may signal I am unwilling or unable to face my irritations, as well as that I am rashing and agonizing over what is the right thing to do.

These simple questions can lead me to profound understanding of myself in life. Insert my symptom and its location in the logical places:

- Am I irritated?
- Am I unwilling to face my irritation?
- Am I rashing over the right thing to do?

With this understanding I can replace my BATHWAVEs to make a real change in the whole system.

To return to the rashing example, perhaps I shift problems with my boss/supervisor and the rash clears up. When I continue these rashing attitudes about my life, work, my purpose, my relationship, my feelings of being unsupported and unfairly judged, or my own judgments of myself, I may out-picture a chronic illness such as asthma or bronchitus. Is that what I want to continue?

The "Descriptionary* offers me questions to understand myself 'mo bettah' as expressed in Hawaii. The attitudes and beliefs are not really causing, nor will they cure, the pain and illnesses. They ARE the condition. The pain and illnesses are simultaneously revealing to me how I am living my life in an out-picture which certainly includes my BATHWAVEs among other things.

The out-picturing of pain and illness is not a judgment of how I am living my life. My life in-picturings and symptoms are the same thing. I am not being judged.

Pain and illness are entanglements.

With the correct constituents, pain and illnesses are always emergent and cannot be withheld, but those constituents do not cause the ailment, they are the ailment. They are the sicknessing, the paining, the asthma'ing, the rashing.

*There is a website http://www.wisdomofthebody.net/index.html that may give me more detail. A nurse practitioner/midwife Constance Smith is currently writing a book that is soon to be published with great detail and examples.

Say my quantum replacements and abidingself sentence once out loud with my eyes closed.

DAY 4
KNEE-D TO KNOW HOW LOW I CAN GO...FLEXIBILITY

MORNING PRACTICE: Say my QC & R.

Knee Joint - Flexibility *Flexibility moving forward into life*	
Love Letter Example	Swollen arthritic knee
Possible Meaning	I may be too emotional about my judgements moving forward in life.
Part 1: **Face & Embrace**	I face and embrace that I am judging myself as I move forward in life and am fearful about my future.
Part 2: **Replace with Grace**	I can, I will and I am living life with confidence, flexibility and without judging in this moment.

EVENING PRACTICE:

Ankle Joint - Flexibility *Connecting my motives (lower legs) and my new understandings (feet)*	
Love Letter Example	Sprained Ankle, swelling and pain on the outer side (present time). Swelling is related to emotions and pain is related to judging.
Possible Meaning	I may be inflexible, feeling overextended or overwhelmed. I am having a hard time connecting my motives (lower legs) and my new understandings (feet)
Part 1: **Face & Embrace**	I face and embrace that I am feeling scared about all of the new understanding I am gaining and overwhelmed.
Part 2: **Replace with Grace**	I can, I will and I am excited and open to living my new understandings.

Say my quantum replacements and abidingself sentence once out loud with my eyes closed.

DAY 5
EMPOWERING MY LIVING

MORNING PRACTICE: Say my QC & R.

Hips *Putting self-empowerment into action*	
Love Letter Example	Pain in both hips.
Possible Meaning	I may be hesitant to put my personal power into action. I am lacking self-confidence in my gifts, feeling incompetent and powerless. I immobilize my actions, my movements.
Part 1: **Face & Embrace**	I face and embrace I give away my personal power to others.
Part 2: **Replace with Grace**	I can, I will and I am confident living my personal power.

Lower Back/Kidneys *The essence of the kidney is openness, trust, and gentleness*	
Love Letter Example	Pain in lower back (location of the kidneys).

Possible Meaning	The back represents the past and pain would indicate judgements. I may not trust someone from my past or myself, going back on a promise. Am I closed down, backing away, or abandoning someone or something? Are my words harsh, abrasive, or contradictory to myself or others? I worry about failing.
Part 1: **Face & Embrace**	I face and embrace I am not trusting others from my past feeling insecure and spiteful.
Part 2: **Replace with Grace**	I can, I will and I am trusting, secure, and gentle with myself first and others as I desire. **Remember there are no failures** I face and embrace that I believe failure exists. I can, I will and I know failure does not exist.

EVENING PRACTICE:

Shoulders Joint - Flexibility *Feeling Burdened*	
Love Letter Example	Right Shoulder
Possible Meaning	The right side of the body is about receiving. Pain in the right shoulder may indicate burdens about what I am receiving in my life with my job, family, or people in my life. It may also be about the burden of doing it right, carrying around commitments, duties/responsibilities.
Part 1: **Face & Embrace**	I face and embrace I'm tired of my old self and I'm ready to be my new self, my resplendent illuminatedbeing.
Part 2: **Replace with Grace**	I can, I will and I do acknowledge my new self that is resplendent, abiding and illuminated.
If pain persists, see if I believe in failure. **Remember there are no failures** I face and embrace that I believe failure exists. I can, I will and I know failure does not exist.	

Say my quantum replacements and abidingself sentence once out loud with my eyes closed.

DAY 6
DREAM IT..LIVE IT

MORNING PRACTICE: Say my QC & R.

Neck *The connection between my thinking and my body (the doings of your life)*	
Love Letter Example	Pain in the back of neck.
Possible Meaning	My neck connects my thinking and the doings of my life. The neck is about living your dreams. Depending on where the pain in the neck is. What from my past keeps me from living my dreams?
Part 1: **Face & Embrace**	I face and embrace I am not smart enough to do a new job.
Part 2: **Replace with Grace**	I can, I will and I am able to do whatever I practice.

EVENING PRACTICE:

Head *I think I can't do it*	
Love Letter Example	Frontal headache
Possible Meaning	Front is about the future. I may think I cannot do what I have to do today or in the coming days.
Part 1: **Face & Embrace**	I face and embrace cannot make dinner for my family.
Part 2: **Replace with Grace**	I can, I will and I am doing all I need.

Say my quantum replacements and abidingself sentence once out loud with my eyes closed.

DAY 7
IN MY GIVING IS MY RECEIVING, IN MY RECEIVING IS MY GIVING

MORNING PRACTICE: Say my QC & R.

Wrist Joint - Flexibility/Giving/Receiving	
Love Letter Example	Pain in the left wrist on the left side of wrist. The left side indicates what I am receiving. Pain indicates judgements.

Possible Meaning	Am I inflexible, not able to bend in what I give/receive.. Or possibly too fussy or picky. I may be demanding and stubborn. Am I holding onto something or am I restricting myself from what I receive?
Part 1: **Face & Embrace**	I face and embrace that I am holding onto what I receive, I do not want to fail and be judged by others.
Part 2: **Replace with Grace**	I can, I will and I am receiving my gifts without judging.

EVENING PRACTICE:

Jaw Communicating convictions	
Love Letter Example	Tightness, clicking, or misalignment in jaw
Possible Meaning	The location of tightness/pain is on the side of my body meaning present time. Bones relate to one's convictions. My mouth communicates with others. I may be restricting my communication to others about my convictions.
Part 1: **Face & Embrace**	I face and embrace that I resist communicating my convictions with others.
Part 2: **Replace with Grace**	I can, I will and I am open to communicating my convictions with others while being more relaxed.

Say my quantum replacements and abidingself sentence once out loud with my eyes closed.

(The Five Elemental Relationships)
Resplendency is Elementally EEEZY!

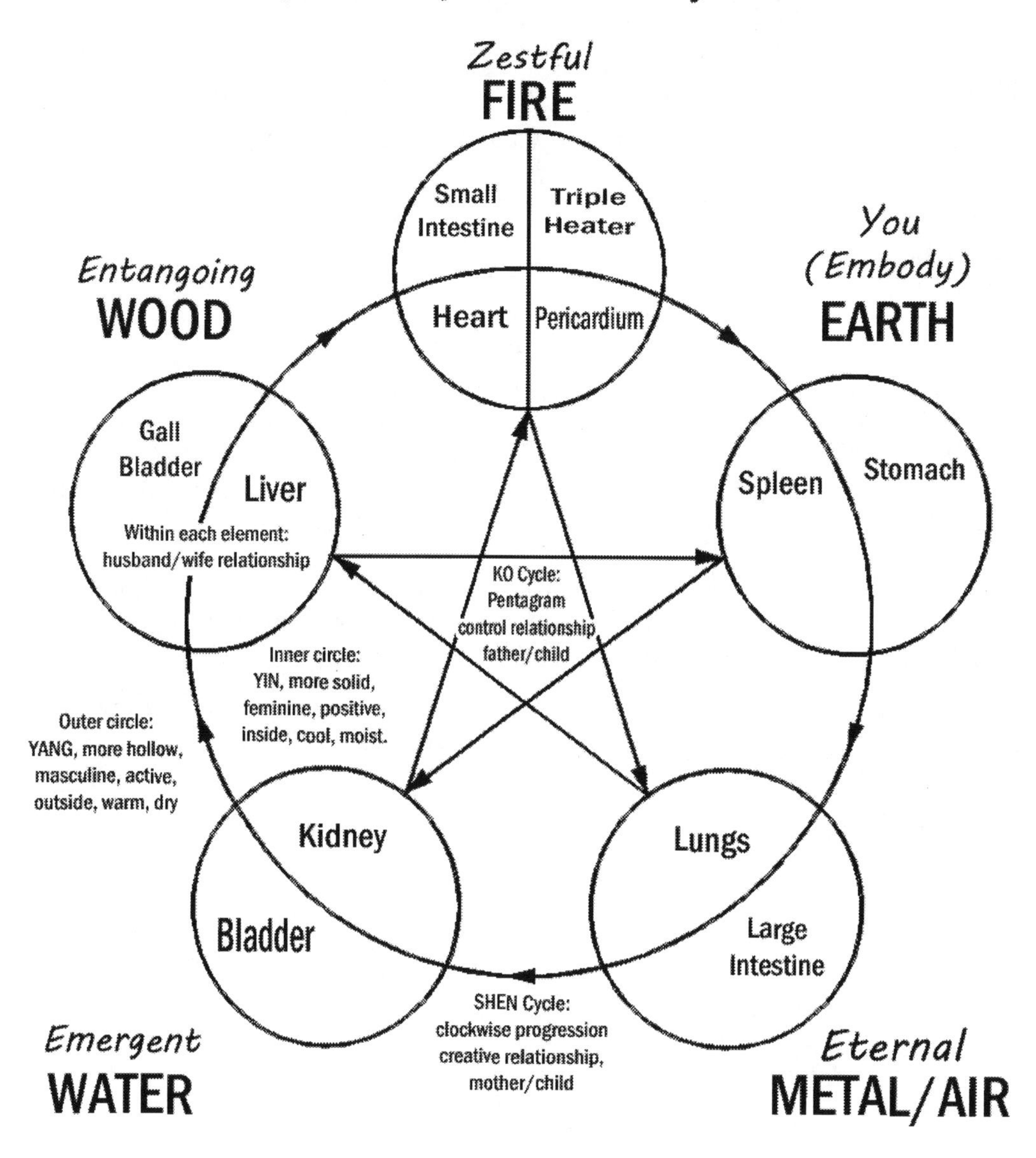

Eternal Metal/Air nurtures Water
Emergent Water nurtures Wood
Entangoing Wood nurtures Fire
Zestful Fire nurtures Earth
You embody Earth nurturing Metal/Air

DAY 8
FIVE ELEMENTS OR WISDOM OF THE ENERGY

MORNING PRACTICE: Say my QC & R.
To better understand how to read the love letters that I am giving myself with my body, it helps to know about the relationships between each of the main body elemental organs. These five elements are all in multiple natural relationships with each other.
As part of my Traditional Chinese Medicine training, knowing about the five element flow of social relationships has never ceased to amaze me in its efficacy and accuracy. In support of Relationshifting, I have noticed that the physiological organ relationships reflect these social relationships in my life like our Relationshifting mirror-calls. This is consistent with wholographic theory presented earlier.

EVENING PRACTICE:
I am only using the 5 Element Relationshifting method to encourage the following:

"You can only regenerate the body directly in two ways:

1). Do only those things you are willing to do eternally and have done to you eternally.

2). Do only those actions you are proud of doing." Cotting, R. B. (2012)

Ask for dreams about how I can do this.
Say my quantum replacements and abidingself sentence once out loud with my eyes closed.

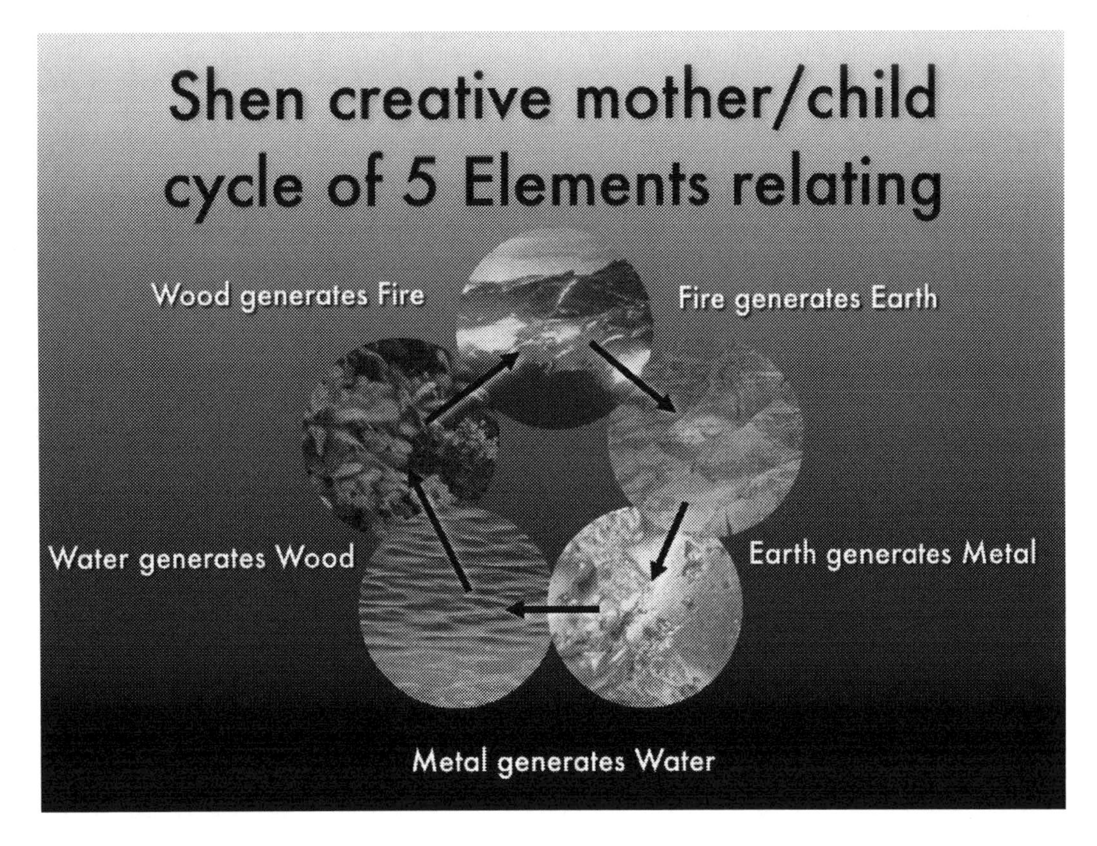

DAY 9
FIVE ELEMENTS FAMILIA

MORNING PRACTICE: SHEN OR MOTHER/CHILD CYCLE

Say my quantumcode & replacements.

To begin with, the circle called the Shen or Chen cycle of the Mother/Child Law tells me about the supportive, nurturing and creative relationship of a mother and a child.

When I start at the top of the circle with the **fire** element, I know that as fire burns, it generates ashes and therefore, **earth**. Therefore fire would be the mother of earth.

As earth compresses, it forms **metal** and **air**, which is its child. The metal and gases inside the earth such as hydrogen and oxygen interact, becoming its child, **water**.

Moving around the circle, water as it falls on the earth's seeds, gives birth to its child, **wood** and the plants. When we strike the wood against itself, it gives birth to its child, **fire**.

I have created a full circle of these five elements and the Mother/Child relationships. Study these if I wish to understand a lot of traditional chinese medicine. Draw them and even color them if I like. Tonight I will learn the Father/Child relationships to each element.

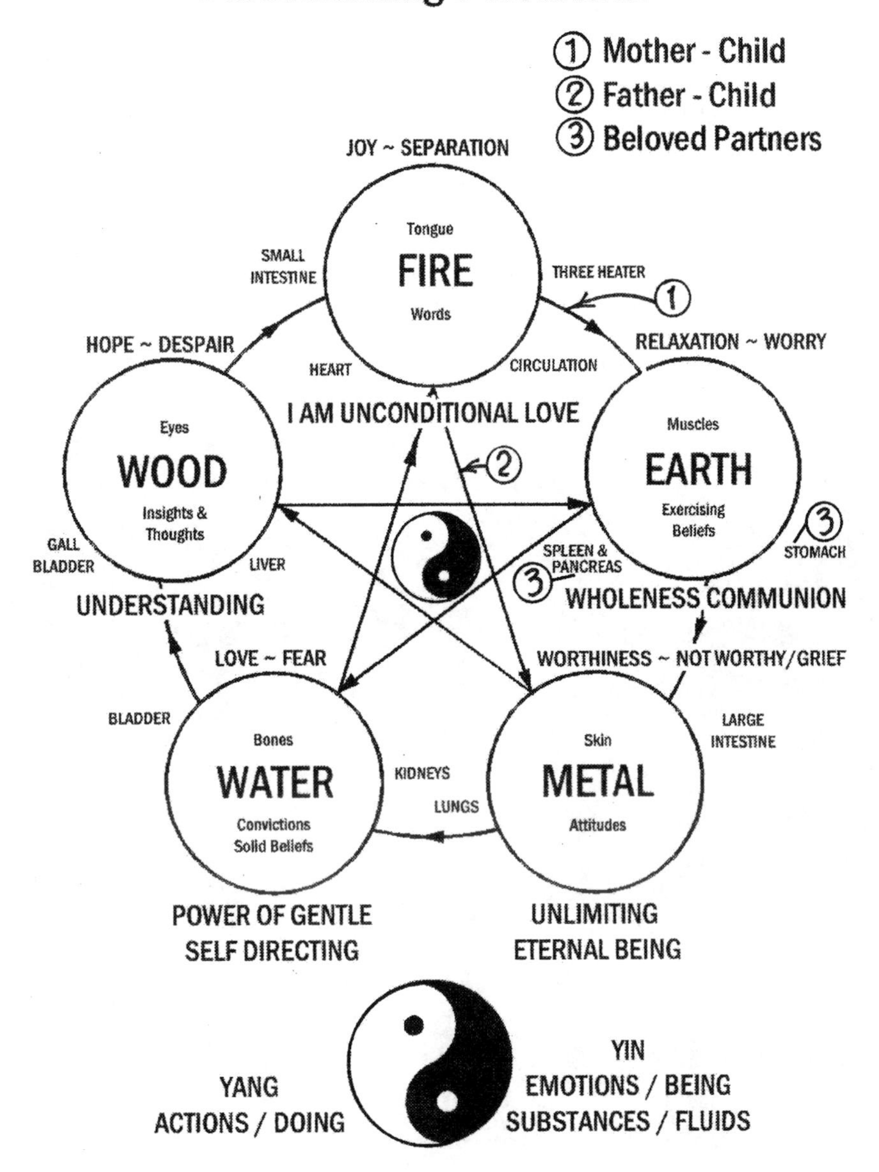

EVENING PRACTICE: GUIDANCE AND DISCIPLING STAR CYCLE

Within the circle, I can see the form of a star, which is the guiding or disciplining cycle. As long as I can draw a 5-pointed star without picking up my pen, this will be easy. The direction my pen flows as I connect the elements, gives me the direction of the fathering.

For instance, the child, **fire**, is cooled by its father, **water**.

Water is then contained and guided by its father, **earth**.

The **earth**-child is held together by its father, **wood's** roots.

The **wood**-child is pruned by its father, **metal**, which is the substance which cuts wood.

The **metal**-child is melted by its father, **fire** and we're back again to the **fire**-child, cooled by its father, **water**.

These cycles are very easy to remember and are useful to remember since all the organs relate to each other within this framework.

Draw the Father/Child - Guidance and Discipling Star Cycle.

Say my quantum replacements and abidingself sentence once out loud with my eyes closed.

DAY 10
THREE RELATIONSHIP CYCLES INCLUDED IN FIVE ELEMENTS

MORNING PRACTICE: THE BELOVED CYCLE WITHIN THE ELEMENTAL ORGANS

Say my quantumcode & replacements.

Within each element, there is a beloved partnership between a yin and a yang organ. From the oriental point of view, yin and yang are the basic phenomena of the dance of the two basic forces of energy in the universe.

Yin is described as 'feminine' and receptive. It is associated with the qualities of coolness, wetness, darkness, contraction, nighttime, and earth. In the body, yin manifests as all biological fluids including blood, saliva, cerebral spinal fluids, semen, etc.. Yin symptoms may include low blood pressure, pale color, chronic conditions and the parasympathetic (relaxing nervous system), to mention a few. From the quantum point of view it is simply 'opening sacred space'.

Yang, on the other hand, is described as the 'masculine' energy. It is demonstrated by heat, dryness, light, projection/expansion, day and heaven. Yang energy manifests in the body as fevers, energy, the digestive and immune systems, acute conditions, high blood pressure, redness in any part of the body and the sympathetic central nervous system, which is the fight or flight aspect. From the quantum point of view, it might be 'expanding active focus'.

Understanding that these two forces are always dancing polarities, and not opposites, is a really important distinction from Western ideas of these qualities. In Chinese medicine, these forces are beloveds. Yin and yang support one another. They are absolutely necessary for each other, they birth each other, as characterized by the popular yin yang symbol shown in the figure.

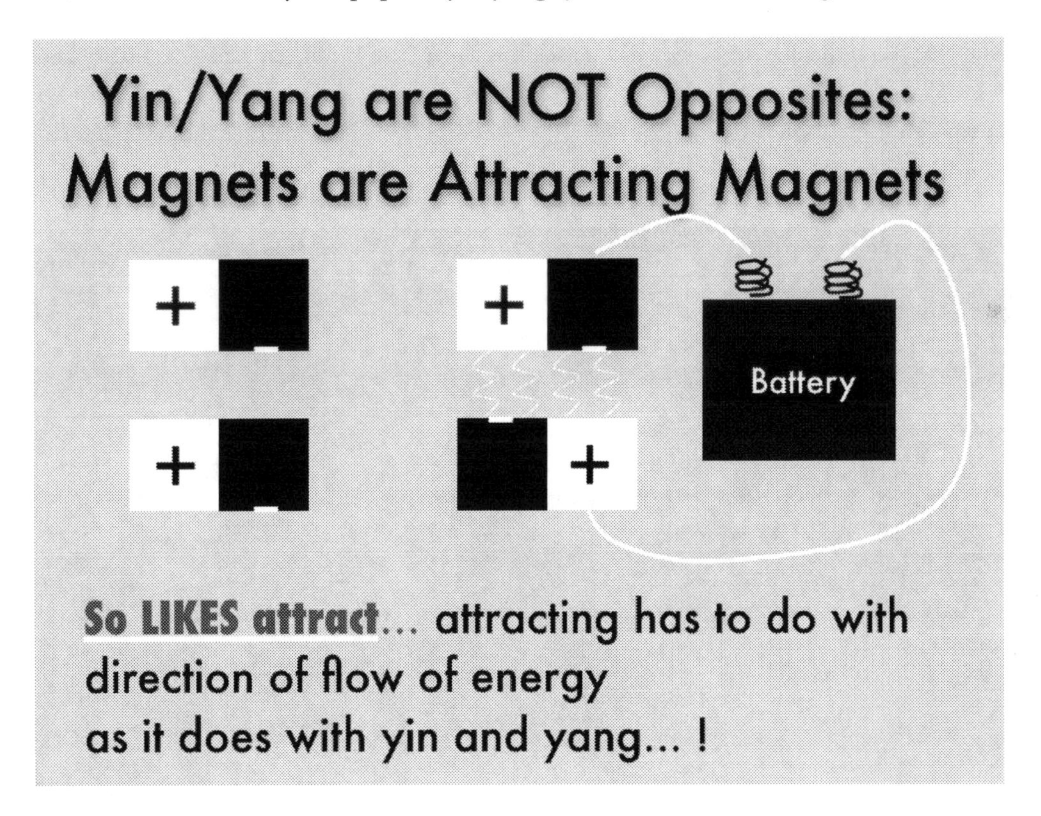

EVENING PRACTICE: CHERISHING BELOVED CYCLE

Within each element there is a cherishing beloved cycle of organs. Their gender can be remembered with a simple rule that the yin organs are generally more solid as compared to their yang partners that appear hollow.

Remember that the beloveds are mirror-calls of our self-love!

Within red **fire**, there is a partnership between the yin heart organ and the yang small intestine organ. Yin written inside and yang outside the large Creative cycle in the figure.

Earth's yellow/golden partnership on the inside of the big creative cycle is the yin spleen/pancreas system, which is partnered to the yang stomach organ.

Within silver/white is **metal/air's** partnership. The yin lung organ is partnered to the yang large intestine organ.

Within blue **water's** partnership, we find the yin kidney married to or partnered to the yang bladder organ, written outside the creative circle.

In the green **wood** partnership, the yin organ is liver on the inside of the mother-child cycle. On the outside, we see the yang organ, gallbladder. These partnerships are flowing within each element.

This whole five element picture, if it could be seen with its 'five color' energies in motion, would be quite dynamic. Seen as a swirling rainbow circle and star echoing the personal and abidingself. These arrows in the diagram show a motion or glowing of energy that is constantly happening within our dancing bodily systems.

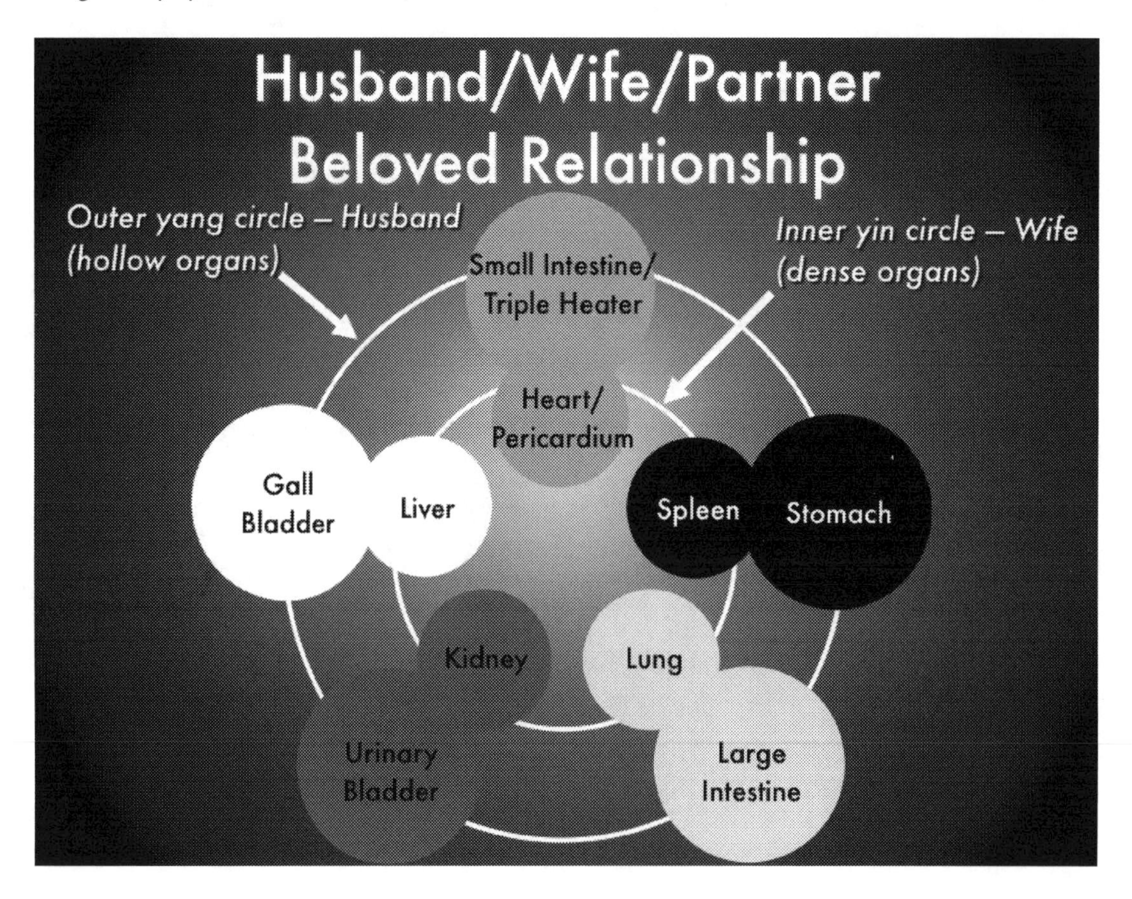

Draw the cherishing beloved cycles. These cycles will be mine forever if I draw them once or twice a day for a week.

Say my quantum replacements and abidingself sentence once out loud with my eyes closed.

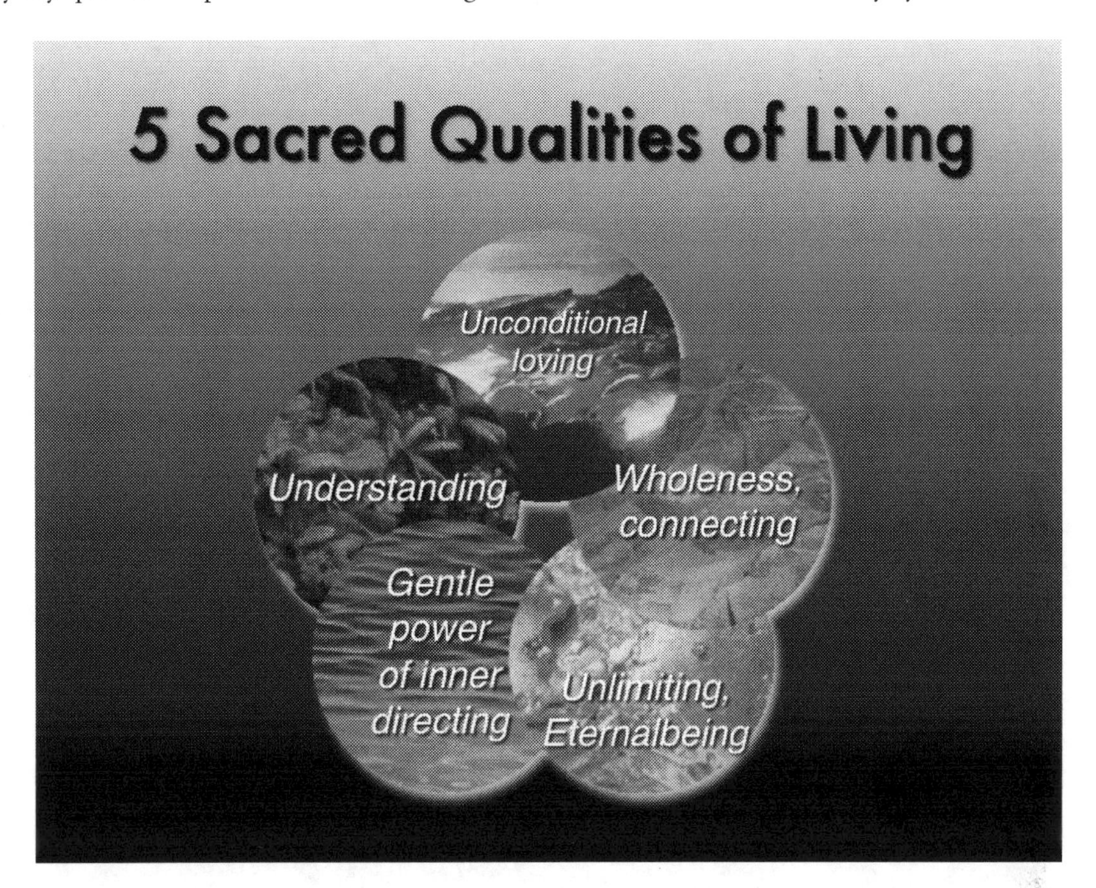

QUANTUM REPLACEMENTS FOR MY BODY

I will repeat these present tense replacements for at least 21 days to 3 months (highly recommended). Before bed say with eyes closed. In the morning say with eyes open.

I am living life with confidence, flexibility and without judging in this moment.

I am excited and open to living my new understandings.

I am confident living my personal power.

I am trusting, secure, and gentle with myself first and others as I desire.

I do acknowledge my new self that is resplendent, abiding and illuminated.

I know failure does not exist.

I am open and able to do whatever I practice.

I am doing all I need.

I am receiving my gifts without judging.

I am open to communicating my convictions with others while being more relaxed.

DECADE 7

Grounding my Earthen Element of Wholeness AS A TRIUNITY

"Mirror, mirror on the wall,
What's the mirror-call of all?"
"Recognize self through your eyeballs."

"Mirror, mirror on the wall,
What makes me feel so small?"
"Your self criticism and judgment call."

"Mirror, mirror on the wall,
How can I get on the ball?"
"Resplendently do as you enthrall."

ESSENCE OF EARTH: GIFT OF EMBODIMENT

As vast as the earth is to my body, that immensity does not touch the vastness of my quantum wave self to my personalself.

That quantum wave self pulsates my radiant unique pattern manifested by the personalself, throughout the whole universe of time and space. I navigate this body as this universal awareness which fulfills and satisfies me; it is pulsating and undulating for all, recorded through entinglement. This is the gift of earthen embodiment.

How can anything happen to or perturb this vastness that is not my doing? I am a master of my vastness. Listening to the quantum wave-like communicating images in dreams, my body, daily events these are my visions. I replace them by replacing my visions with new informotion of my choosing. The additions are actions of emergent surprises in this moment.

This is what living is like a surprise each moment as I desire. I transform the whole universe as my abidingself of my quantum wave pattern.

The Mayans said that after the age of Galactic awareness is Universal awareness.* This is what it means: that being aware responsibly of myselves (my triunity). I am manifesting the universe as I desire, communicating as emergent miracles or surprises in this moment. This expands myself. And through entanglement (entinglement) everyone (myself included) has access to this knowledge of the universe. All knowledge is there for the asking and listening with universal (quantum) ears.

That changes the whole approach to education doesn't it? How exciting finally. Everyone has access to universal knowledge. Knowledge is not to be 'fed' to students as past beliefs that may exist. Empowering student awareness to access universal knowledge that is free for all to receive. Practice exercising the brain/body/selves in asking questions and knowing how to receive responses.

The whole universe is quantum based as quantum is universal.

This is nurturing my embodiment as earth does. The earth appears as the seeming solid element that supports growth of food for my nurturance. Formed and warmed by the fire of its mother lava. Earth's body is composed of years of mineral formation. It supports my embodiment. It serves me without taking credit. Earth is solid in its accountability and integrity.

The food and raw materials it produces fulfill my needs. It reminds me how to fulfill my needs and express myself in satisfying nurturing ways. The earth by its nature is self sustaining. I am composting, renewing, resting, and restoring as earth shows me how to energize myself.

My stomach receives the food of earth and dissolves it with battery acid ph into the components that can be absorbed by my body. The pancreas neutralizes it to digest fat and absorb sugar preparing me to see the sweetness of living energized. The spleen is the largest lymph node showing us the safety of my self protecting and filtering.

Can happiness find a home in me? Can I melt into the lava of living? Can I flow and solidify wherever I land for all of this is the essence of earth.

EARTHEN BODY FROM A QUANTUM PERSPECTIVE

They said everything was changing on earth, but I didn't know it was faster than the speed of light. 10 with 44 zeros after it times per second. We didn't know it was pulsating in and out of existence being new every moment!

PERSONAL EARTHY MUSING: **The Earth is my Drum, my Tomtom**

I just awoke from a dream in which people were squirting each other in the closed eyes of faces with gentle water hoses, in fun. Since water is emotion it means I am waiting on people's emotions to shower me, for feelings to motivate or impact me. I am blinding myself as a junky to good feelings. Feelings are temporary experiences and usually reactions; therefore, I am replacing that with sharing the informotion of my quantum wave pattern with my 'feeling' in everything I do. With the idea that I am creating these states in myself to help make my writing more personal, where people on earth are coming from.

I face and embrace that these **feelings are often reactions to judgings,** that I have been replacing over and over. Maybe that can help me be more real instead of being the rescuing healer?!

There may be an old belief system behind them that I am not good enough or that I have to prove myself. Maybe I am here to please someone else; like my clients, partner, parent or mentor?

I can, I will and I am ok (good enough) for my quantum wave self, which is my eternal abiding nature that I have manifested.

I am being as vast as the universe in my own unique way. It doesn't have to be difficult?

Quantum wave living is not difficult. It is interactive...like play!

I playfully interact with my quantum wave expanding the universe as I am truly desiring and truly living.

When I stop upsetting myself and access new informotion, my resplendentbeing opens up with emergent surprises called miracles. When I am not so addicted to "feeling good" and sometimes, to "feeling bad", I start living radiant. I am interactive with my eternalself, which is creative. Emergent is my radiant vast awareful observing resplendentself with not much feeling, though not numb.

Feeling is usually a reaction of life happening to me instead of me happening to life!!!

When I am happening to life I am too busy to feel. Do I ever notice that?

Living is not about reacting or judging. Living is about what I initiate when I live my quantum wave pattern and open to my new INFORMOTION. Living is interacting and exchanging, which is opening emergent miracle surprises.

Like a middle Eastern poet Rumi wrote, modified in my quantum haiku:

Beyond thoughts of right
And wrong, there is my quantum field.
I'll see ya there!

I meet my whole illuminatedself here, all three selves.

It's a physical place right here in the universe...

From Angela Reweaver of True Triune Communion:

So I am never alone. There are three of my beings. One, that I see, yet it isn't all I am. The second is my vast quantum interactive wave invisible, yet physical partner. Third, is the surprising new and different me resplendently witnessing the interactions of my personal and eternalbeing in sheer delight! She/he is an expanding new miracle of awareness emergent in this moment. **There are two triunities of me. Wow! How can I ever feel lonely? I am also another virtual triunity of my selves as I envision! The two triunities of me together make the six-pointed star, the basis of sacred geometry's flower of life.**

Diagram of the Virtual TriUnity in Communion with Adorned TriUnity

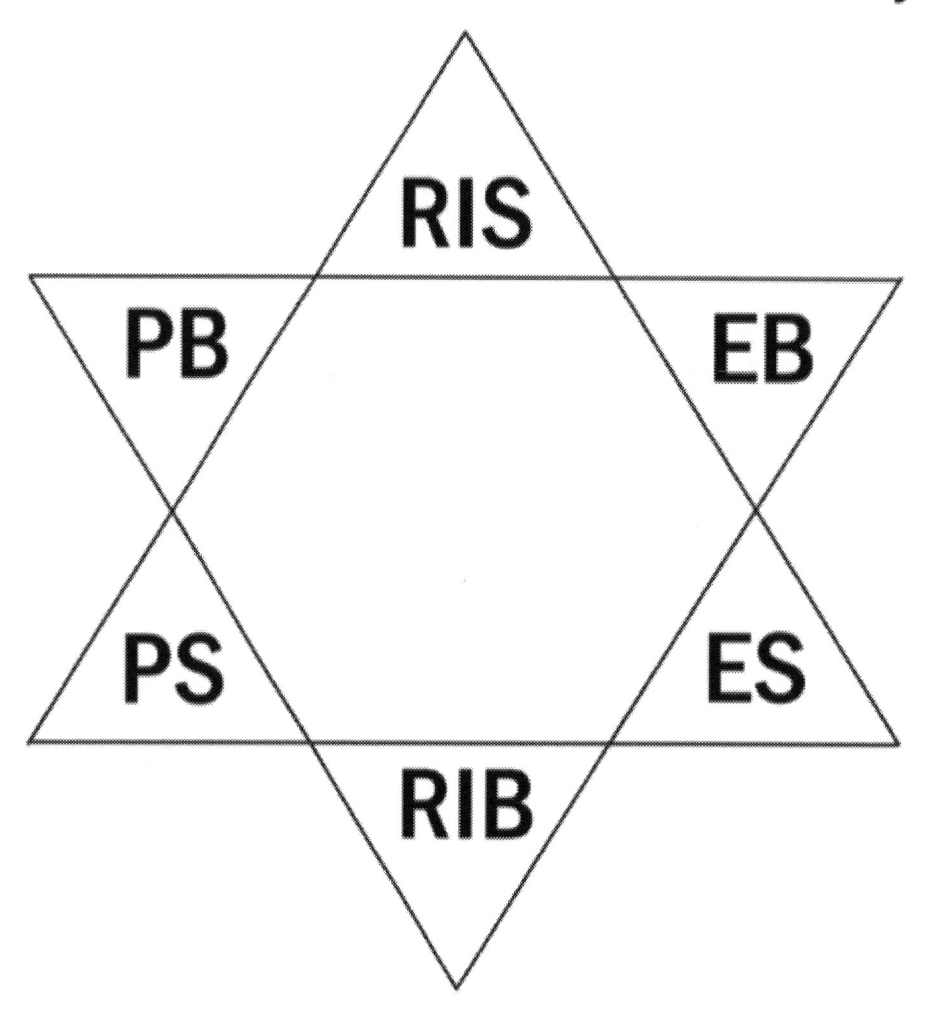

Basic Building Block of Sacred Geometry's Flower of Life

EARTH TO EVERYONE: THIS CHART BELOW IS INFORMOTION FOUND IN MY FIRST BOOK. THE SIMILAR CHARTS OF THE OTHER FOUR ELEMENTS ARE ACCESSIBLE THERE.

YIN/YANG EARTH ORGANS:	**MANIFESTATIONS OF EARTH ENERGY:**
SPLEEN/PANCREAS, the solid organ partner STOMACH, the hollow organ partner **SPLEEN/PANCREAS MOTION:** Spleen is relaxing. Pancreas is sweetening. **STOMACH MOTION:** Stomach is nurturing and satisfying. **SACRED QUALITY OF EARTH:** Networking wholeness. Weaving the web of our living. **OUR EMOTIONS OBSCURING FLOW:** Feeling obsessive, embittered, alienated, isolated, worried, overwhelmed, exhausted, anxious. **OUR EMOTIONS ENHANCING FLOW:** Feeling centering, self nurturing, fulfilling, grounding my desires, relaxing, restoring, rejuvenating, accountable, supportive, renewing.	In body: Lips and muscles. Sense commanded: Tasting. Fluids: Saliva. Bodily smell: Fragrant. Flavor craving: Sweet. Sound of voice: Singsong. Weather that bothers you: Damp. Season you like: Mid summer. Color of skin: Yellow. **RELATIONSHIPS TO ORGANS:** Spleen is the child of the mother heart. Spleen is the child of the father liver. Spleen/pancreas is mother to the child lungs. Spleen/pancreas is the father to the child kidney. Stomach is the child of the mother small intestine. Stomach is the child of the father Gall bladder Stomach is the mother of the large intestine. Stomach is the father of the bladder.

DAY 1
QUANTUM WAVE SMILING PRACTICE

MORNING PRACTICE: Say QC (quantumcode) & R (replacements) eyes open.

BATHWAVE TIME FOR STOMACH BELIEFS:	
BATHWAVEs **Part 1: Face & Embrace**	**Quantum Shifts** **Part 2: Down my Midline with Hands I Replace with Grace**
I face and embrace in my whole body that I worry and I am ready to replace worry.	I can, I will and I am relaxing and trusting myself and appreciating all my mirror-calls.

Pour this freshness into my day.

EVENING RELAXING:

BATHWAVE TIME FOR STOMACH ACTIONS:	
BATHWAVEs **Part 1: Face & Embrace**	**Quantum Shifts** **Part 2: Down my Midline with Hands I Replace with Grace**
I face and embrace in my whole body that I am tired for the turmoil.	I can, I will and I am energizing and enlivening doing my desires.

Read the quantum replacements through and then do the visioning: *Quantum Wave Smiling.* I can record it on a device and listen while doing. Before saying my replacements, address my abidingself by reciting my reading from Decade 2.

QUANTUM WAVE SMILING PRACTICE

Picture my outgoing quantum waves as unbreakable bubbles concentric with real big smiling faces ;-) on each one. These bubbles are expanding throughout the universe embracing billions of galaxies like my own Milky Way Galaxy. Everyone in their solar systems are feeling that embracement, that expansion.

As I approach the edge of the universe I bounce back again. I am connecting and communicating, gathering new informotion as I return to my wave bubble's center. This is where I spin as a particle of my body again smiling.

Stay there for a moment inside my atoms, seeing all the 'quantum wavicles' smiling. The 3 tiny quarks inside my protons, 'up, up, and down quarks' and inside my neutrons: 'down, down and up quarks'. All are in my atomic nuclei smiling and briefly spinning. The electrons are orbiting around the nucleus spinning, all smiling. I feel this throughout my whole body.

Then after a few spins I expand out again. Repeat 2-3 more times from top 'picture' above, as fast or as slow as I desire. End with a picture of all quantum waves smiling along with the spinning center 'particles'. I may bring up this picture whenever I desire.*

DAY 2
I AM "ANEW" SPIN HAIKU

My incoming quantum
Wave spins matter anew
Whatever I do!

MORNING REPLACING: Say QC & R eyes open.

BATHWAVE TIME FOR STOMACH THOUGHTS:	
BATHWAVEs **Part 1: Face & Embrace**	**Quantum Shifts** **Part 2: Down my Midline with Hands I Replace with Grace**
I face and embrace in my whole body that I am anxious.	I can, I will and I am safe, secure, and fulfilled.

BATHWAVE TIME FOR STOMACH & PANCREAS THOUGHTS:	
BATHWAVEs **Part 1: Face & Embrace**	**Quantum Shifts** **Part 2: Down my Midline with Hands I Replace with Grace**
I face and embrace in my whole body that I think living is not sweet.	I can, I will and I think living is sweet.

EVENING REPLACINGS:

BATHWAVE TIME FOR STOMACH HABITS:	
BATHWAVEs **Part 1: Face & Embrace**	**Quantum Shifts** **Part 2: Down my Midline with Hands I Replace with Grace**
I face and embrace in my whole body that I have an old habit of believing that things don't work out for me and I am replacing that.	I can, I will and I am grateful that the universe nurtures me with abundant gifts.

Repeat the quantum wave smile from yesterday. And say quantum replacements.

DAY 3
STOMACHING MY WORDS AND ATTITUDES

MORNING REPLACING: Say my abidingself and all quantum replacements out loud.

BATHWAVE TIME FOR STOMACH WORDS:	
BATHWAVEs **Part 1: Face & Embrace**	**Quantum Shifts** **Part 2: Down my Midline with Hands I Replace with Grace**
I face and embrace in my whole body that what I say doesn't make a difference.	I can, I will and I am worthy and appreciated by me.
I face and embrace in my whole body that I do not value my own words.	I can, I will and I express myself in satisfying and fulfilling ways.

EVENING REPLACINGS:

BATHWAVE TIME FOR STOMACH ATTITUDES:	
BATHWAVEs **Part 1: Face & Embrace**	**Quantum Shifts** **Part 2: Down my Midline with Hands I Replace with Grace**
I face and embrace in my whole body that I can't focus on myself and I am replacing that.	I can, I will and I believe in and focus on myself and others as I desire.
I face and embrace that I discontent myself.	I can, I will and I stop discontenting myself.

Repeat the quantum wave smile from Day 1. And say quantumcode and replacements.

DAY 4
YUMMY TUMMY VALUES AND EMOTIONS

QUANTUM WAVEY GRAVEY HAIKU

Space is quantum wavey
Just like quantum gravey spins
My own biscuits!

MORNING REPLACING: Say QC & R eyes open.

BATHWAVE TIME FOR STOMACH VALUES:	
BATHWAVEs **Part 1: Face & Embrace**	**Quantum Shifts** **Part 2: Down my Midline with Hands I Replace with Grace**
I face and embrace in my whole body that I am disappointed that I don't have enough.	I can, I will and I am grateful for the abundance that I have.
I face and embrace in my whole body that I don't value myself or others.	I can, I will and I value myself and all others in my relationships.

EVENING REPLACINGS:

BATHWAVE TIME FOR STOMACH EMOTIONS:	
BATHWAVEs **Part 1: Face & Embrace**	**Quantum Shifts** **Part 2: Down my Midline with Hands I Replace with Grace**
I face and embrace that I am depressed and not truly satisfied in what I do.	I can, I will and I believe in what I do and am enthusiastic about living.
I face and embrace that I am obsessed with things and disconnected with living.	I can, I will and I focus on what I desire staying connected to my whole self.

I face and embrace I am dissatisfied and feel sorry for myself and I am replacing that.	I can, I will and I am energized, confident and appreciate all I manifest.
I face and embrace that I am U centered and ungrounded and I am replacing this.	I can, I will and I understand that the universe is benevolent giving me what I do in entanglement.
I face and embrace that I believe that this is too good to be true.	I can, I will and I see my abidingbeing imaged in everything I see and do, think, feel and dream.

And say quantum replacements revealing my abiding self out loud.
Repeat the quantum wave smile from Day 1.

DAY 5
BEGIN AGAIN WITH SPLEEN AND PANCREAS PATTERNS

MORNING PRACTICE: Say my abidingself and all quantum replacements out loud.

BATHWAVE TIME FOR SPLEEN/PANCREAS BELIEFS:	
BATHWAVEs **Part 1: Face & Embrace**	**Quantum Shifts** **Part 2: Down my Midline with Hands I Replace with Grace**
I face and embrace that I believe I am alienated or isolate myself from people and I am replacing that.	I can, I will and I believe I interact, connecting with people in many ways like entinglement.
I face and embrace that I believe things happen to me and to people and I am replacing this.	I can, I will and I am happening to my living.
I face and embrace in my whole body that I am a stranger to myself and I am replacing that.	I can, I will and I am aware of everything about me.
I face and embrace that I am hard on myself and judging myself and I am replacing this.	I can, I will and I am gentle and unconditionally loving myself.

EVENING REPLACINGS:

BATHWAVE TIME FOR SPLEEN/PANCREAS ACTIONS	
BATHWAVEs **Part 1: Face & Embrace**	**Quantum Shifts** **Part 2: Down my Midline with Hands I Replace with Grace**
I face and embrace in my whole body that I am withdrawn and I am replacing that.	I can, I will and I am open to reaching out and connecting.
I face and embrace in my whole body that I react and I am replacing this.	I can, I will and I initiate and originate my living.

I face and embrace that I isolate and divide myself into parts and I am replacing that.	I can, I will and I am whole and authentically me.

DAY 6
OPENING SPLEEN/PANCREAS THOUGHTS AND HABITS

SPACEY HAIKU

Space is opening.
Open is sacred. Space is
Sacred quantum waves.

MORNING REPLACING: Say QC & R eyes open.

BATHWAVE TIME FOR SPLEEN/PANCREAS THOUGHTS:	
BATHWAVEs **Part 1: Face & Embrace**	**Quantum Shifts** **Part 2: Down my Midline with Hands I Replace with Grace**
I face and embrace in my whole body that I am exhausting myself.	I can, I will and I energize myself with everything I do.
I face and embrace that I feel overwhelmed.	I can, I will and I focus on and dedicate myself to those things I desire.

EVENING REPLACINGS:

BATHWAVE TIME FOR SPLEEN/PANCREAS HABITS:	
BATHWAVEs **Part 1: Face & Embrace**	**Quantum Shifts** **Part 2: Down my Midline with Hands I Replace with Grace**
I face and embrace that I have a habit of needing to worry to live and I am replacing that.	I can, I will and I relax while I do what I desire.
I face and embrace in my whole body that my old habits are repeating.	I can, I will and I am in charge of my living and replace all habits not authentically me.

Repeat the quantum wave smile from day 1. And say quantum code and replacements.

DAY 7
SPLEEN/PANCREAS WORDS AND ATTITUDES OF WHOLENESS

SELFNESS HAIKU

This is THE AGE of
'SELFNESS', as I live RESPLENDENT-
BEING AWARE.

MORNING REPLACING: Say QC & R eyes open.

BATHWAVE TIME FOR SPLEEN/PANCREAS WORDS:	
BATHWAVEs **Part 1: Face & Embrace**	**Quantum Shifts** **Part 2: Down my Midline with Hands I Replace with Grace**
I face and embrace in my whole body that my words separate, divide and are nervous.	I can, I will and I am expressing in satisfying ways as my whole resplendent illuminatedbeing.

EVENING REPLACINGS:

BATHWAVE TIME FOR SPLEEN/PANCREAS ATTITUDES:	
BATHWAVEs **Part 1: Face & Embrace**	**Quantum Shifts** **Part 2: Down my Midline with Hands I Replace with Grace**
I face and embrace in my whole body that I live in the environment that is obsessive and worrying and I am replacing that.	I can, I will and I live in a self sustaining supportive environment.
I face and embrace in my whole body that I am a loner unwillingly and I am replacing this.	I can, I will and I realize that as my triunity (my) personal, abiding and resplendent (self/unity), I am never alone reaching the whole universe without effort.
I face and embrace in my whole body that I am disconnected and I am replacing that.	I can, I will and I am in entinglement always with everything so I easily connect.
I face and embrace that I breathe shallowly.	I can, I will and I breath deeply in this moment.

Repeat the quantum wave smile from Day 1. And say all quantum replacements out loud.

DAY 8
REPLACING SPLEEN/PANCREAS VALUES AND EMOTION PATTERNS

MORNING REPLACING: Say QC & R eyes open.

BATHWAVE TIME FOR SPLEEN/PANCREAS VALUES:	
BATHWAVEs **Part 1: Face & Embrace**	**Quantum Shifts** **Part 2: Down my Midline with Hands I Replace with Grace**
I face and embrace that I separate my body/ mind, quantum wave pattern/abidingself and resplendent illuminatedself.	I can, I will and I live my whole self without hierarchy or judging.
I face and embrace in my whole body that I separate myself into two parts and then make up names for the two parts like masculine and feminine and I am replacing this.	I can, I will and I am a whole functioning personalbeing living in a partnership with a whole unique abidingbeing. As a partnership we are manifesting new resplendent illuminatedbeing faster than we can imagine expanding the universe.

EVENING REPLACINGS:

BATHWAVE TIME FOR SPLEEN/PANCREAS EMOTIONS:	
BATHWAVEs **Part 1: Face & Embrace**	**Quantum Shifts** **Part 2: Down my Midline with Hands I Replace with Grace**
I face and embrace that I am sacrificing or martyring my self and I am replacing this.	I can, I will and I enliven myself.
I face and embrace in my whole body that I hurt myself and I am replacing this.	I can, I will and I do invigorate and restore my well being.
I face and embrace in my whole body that I feel alienated or disoriented and I am replacing that.	I can, I will and I am on track and in my orbit connecting as I desire.
I face and embrace in my whole body that I feel stressed, worried and anxious.	I can, I will and I realize that 'stressed' spelled backwards is desserts, so I am enjoying living as a bowl full of cherries.(or box of chocolates.)
I face and embrace that I feel edgy and panicky and I am replacing that in this moment.	I can, I will and I am rested, refreshed and restored.
I face and embrace in my whole body that I feel I am obsessing and fatiguing myself.	I can, I will and I expand opening to the big picture and feel energized.

Repeat the quantum wave smile from Day 1. And say my abidingself and all quantum replacements out loud.

DAY 9
EMBODYING ME AS I DESIRE TO EARTHEN LIVING

Quantum Wave Mirror-calls Haiku

Your wave meets my wave;
Mattering begins to
Interact miracles.

MORNING PRACTICE: Say my abidingself and all quantum replacements out loud.

Congratulations on completing freeing up earthen relationships (elements) for the world and googleverse.

It's time to make a recording of all of my EARTHEN elemental replacements.
Search today for relaxing yet inspiring music without words that I can use as background music for recording my reading tonight. Possibly from Pandora. It may take 10-20 minutes.

If I have neither, no worries. I can imagine hearing the ocean shore or the birds singing.

Smart phones and/or computers have recording apps. If mine doesn't, I can download one for free. Get help if I don't know how to do that. Sometimes I will find the app in utilities or tools.
If I have no smart phone or computer or recording device, then I will read them from the end of the chapter out loud. It will take less time since I only read them once. If I am a visual learner, reading may be my best method to use.

EVENING PRACTICE:
Find my abidingself reading. Begin and end my recording reading it with spaces so I can repeat it outloud. Also gather the replacements from dreamwaving (last Decade) and relationshifting before that. Do them all together until you feel they are you! (minimum three weeks, sometimes takes three months to live them)

When recording, I'll leave a silent space with just music between each replacement so that I can repeat them out loud when listening. The recording makes it easy to keep my eyes closed at night and open in the morning.

When I am ready to record, turn on the music and press record. Take deep breath before reading each replacement. Read them PASSIONATELY yet gently as if I am reading a love letter.

DAY 10
MOVE EXPRESSLY ON THE EARTH

EARTHRISE HAIKU

As I walk I feel
The earth rise up to greet my feet;
It moves me on.

MORNING PRACTICE: SUSCITATE MYSELF WHILE WALKING
Be sure to record my dreams first and do any replacements I wish from that. Write them down to be read. Play the recording I made last night repeating out loud or read them as such.

When I am up to it, do spontaneous movements while listening and speaking. It may be as simple as swinging my arms front to back. This is actually a form of chi gong for energizing the immune system. Cup my hands slightly as they swing. Any other movements are wonderful.
Live these replacements today as I desire.

EVENING OBSERVATIONS:
The quantum world reveals that the universe is such that it supports my illuminatedbeing. For me it doesn't get any better than that.
I am living nurturing, fulfilling, and satisfying days. Write automatically about that and anything else I believe.
Play the recording I made last night repeating out loud or read them, closing my eyes to speak them.
In the next Decade the pedal is put to the metal patterns of my great value.
The universe appreciates my persistent commitment!

QUANTUM REPLACEMENTS FOR MY EARTH ORGAN STOMACH

I will repeat these present tense replacements for at least 21 days to 3 months (highly recommended). In the morning with eyes open and in the evening with eyes closed.

I am relaxing and trusting myself and appreciating all my mirror-calls.
I am energizing and enlivening doing my desires.
I am safe, secure, and fulfilled.
I think living is sweet.
I am grateful that the universe nurtures me with abundant gifts.
I am worthy and appreciated by me.
I express myself in satisfying and fulfilling ways.
I believe in and focus on my self and others as I desire.
I stop discontenting myself.
I am grateful for the abundance that I have.
I value myself first, then others in my relationships.
I believe in what I do and am enthusiastic about living.
I focus on what I desire staying connected to my whole selves.

I am energized, confident and appreciate all I manifest.
I understand that the universe is benevolent giving me what I do in entanglement.
I see my abidingbeing imaged in everything I see and do, think, feel and dream.

QUANTUM REPLACEMENTS FOR MY EARTH ORGAN SPLEEN/PANCREAS

I believe I interact connecting with people in many ways like entinglement.
I am happening to my living.
I am aware of everything about me.
I am gentle and unconditionally loving myself.
I am open to reaching out and connecting.
I initiate and originate my living.
I am whole and authentically me.
I energize myself with everything I do.
I focus on and dedicate myself to those things I desire.
I relax while I do what I desire.
I am in charge of my living and replace all habits not authentically me.
I am expressing in satisfying ways as my whole resplendent illuminatedbeing.
I live in a self-sustaining supportive environment.
I realize that as my triunity (my) personal, abiding and resplendent (self/ unity), I am never alone reaching the whole universe without effort.
I am in entinglement always with everything so I easily connect.
I breathe deeply in this moment.
I live my whole self without hierarchy or judging.
I am a whole functioning personalbeing living in a partnership with a whole unique abidingbeing. As a partnership we are manifesting new resplendent illuminatedbeing faster than we can imagine expanding the universe.
I enliven myself.
I do invigorate and restore my well-being.
I am on track and in my orbit, connecting as I desire.
I realize that 'stressed' spelled backwards is desserts, so I am enjoying living as a bowl full of cherries.(or box of chocolates.)
I am rested, refreshed and restored.
I expand opening to the big picture and feel energized.

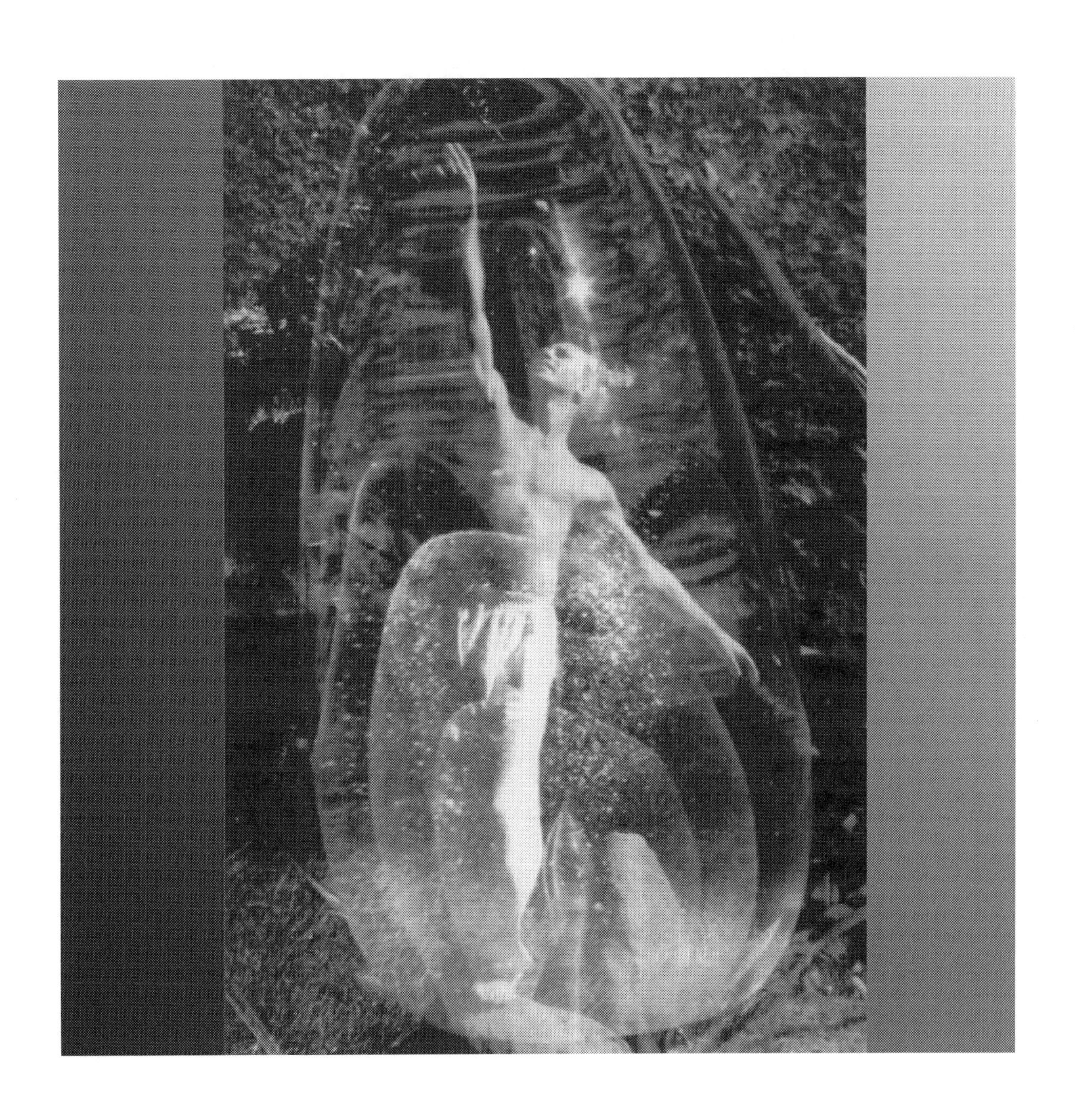

DECADE 8

METAL/AIR ELEMENT
SELF-WORTH

I am taking these 10 days to explore, expand with and transfigure my own metal/air self-worth with my BATHWAVEs!

DAY 1
Essence of Metal/Air: Transfiguring and Valuing Myself

MORNING READING: Say quantumcode and replacements out loud with eyes open.
Compressing without Stressing:
As the earth compresses matter into precious jewels, metal, gases escape from this process. The **lungs concentrate** the oxygen from the air we breathe and convert it into precious chi, energy. **This concentration of metal/air is my moment to moment jewel, the spark, and the treasure of being alive.**
The large intestine fulfills the other essential aspect of metal/air energy, as the culmination of compressing all of my intake and then making space. Of course this completes a cycle for me of taking in and extracting my true desires, compressing and then returning back to my environment for recycling what does not serve me. Both CO2 for plants to turn into oxygen and human waste for plants to use as compost for growth-thus metal guides in fathering the wood element cycle.
EVENING READING: Transfiguring and Valuing Myself
Lung Motion:
Lung is about valuing and energizing my self worth first. Just as in the airplane principle of putting the oxygen mask over my face first before assisting others! Lung is also about valuing the gifts and mirror-calls of others.
Large Intestine Motion:
Large Intestine is about transfiguring my "shit" into "compost". Transfiguring what I don't need to "doodoo" anymore, including my judging of myself and others- anything that doesn't serve me in living what I truly desire.
Large intestine is about making it easy to "move on" from what no longer serves me. Sometimes it's hard to move on and that appears as constipating. I may lose my vital emotions along with the emotions that no longer serve me and find myself diarrhea-ing; wasting the nourishing stuff along with the old.
Sacred Quality of Metal: Valuing the transfiguring of my living in these areas of my BATHWAVEs: **B**eliefs, **A**ctions, **T**houghts, **H**abits, **W**ords, **A**ttitudes, **V**alues, **E**motions below.
Say out loud my quantumcode with my new and old replacements. As I speak, I practice watching a new and different scene of my motion picture with me the star.

DAY 2
MOVING ON FROM WHAT NO LONGER SERVES ME

MORNING PRACTICE: Say quantumcode and replacements out loud with eyes open.

BATHWAVE TIME FOR LARGE INTESTINE BELIEFS:	
BATHWAVEs **Part 1: Face & Embrace**	**Quantum Shifts** **Part 2: Replace with Grace**
I face and embrace that I have to hold on to what I no longer need and I am replacing this.	I can, I will and I joyfully replace the past to make space for change.

Talk about specifics I may hold on to or anything that comes up while embracing it.
Replace it and write it in my notebook.
Take three deep breaths and vision lots of opening sacred space in my living.

EVENING PRACTICE:

BATHWAVE TIME FOR LARGE INTESTINE ACTIONS:	
BATHWAVEs **Part 1: Face & Embrace**	**Quantum Shifts** **Part 2: Replace with Grace**
I face and embrace that I deplete my resources and I am replacing this.	I can, I will and I use my resources benefiting me.

Talk about how I may do anything that comes up while embracing. Replace it, then write the present tense here or in my notebook on my quantum replacings page.
Take three deep breaths and promise myself to take care of self first like with oxygen masks on an airplane .

Say out loud my quantumcode with my new and old replacements. As I speak,
I practice watching new and different scene of my motion picture with me the star.

DAY 3
LARGE INTESTINE THOUGHTS AND HABITS

MORNING PRACTICE: Say quantumcode and replacements out loud with eyes open.

BATHWAVE TIME FOR LARGE INTESTINE THOUGHTS:	
BATHWAVEs **Part 1: Face & Embrace**	**Quantum Shifts** **Part 2: Replace with Grace**
I face and embrace that I block and sabotage my glow and I am replacing this.	I can, I will and I yield to glow with living.

Talk about anything that comes up while embracing it. Replace it writing from the chart in my notebook on my special quantum replacings page.
Take three deep breaths and notice how I feel overall. Have a good day.

EVENING PRACTICE:

BATHWAVE TIME FOR LARGE INTESTINE HABITS:	
BATHWAVEs **Part 1: Face & Embrace**	**Quantum Shifts** **Part 2: Replace with Grace**
I face and embrace that I have to be in control of my own resources and I am replacing this.	I can, I will and I manifest sustainable resources.

Speak about anything that comes up while embracing. Replace it writing from the chart in my notebook or on my special quantum replacings page.
Offer gratitude for my breath and trust my sustainable selves.

Say out loud my quantumcode with my new and old replacements. I practice seeing new and different motion pictures with each one.
Have a good sleep.

DAY 4
GLOWING STRENGTH

MORNING PRACTICE:

BATHWAVE TIME FOR LARGE INTESTINE WORDS:	
BATHWAVEs **Part 1: Face & Embrace**	**Quantum Shifts** **Part 2: Replace with Grace**
I face and embrace that my words are powerless and I am replacing this.	My words can, will and are strong and meaningful.
I face and embrace that I don't keep my promises and I am replacing this.	I can, I will and I keep my promises.

Write about what comes up while embracing it.
Replace it writing from the chart in my notebook on my special quantum replacings page.
Picture yourself this way.

EVENING PRACTICE:

BATHWAVE TIME FOR LARGE INTESTINE ATTITUDES:	
BATHWAVEs **Part 1: Face & Embrace**	**Quantum Shifts** **Part 2: Replace with Grace**
I face and embrace that my life feels stuck and I am replacing this.	I can, I will and I glow freely through living.

Speak anything that comes up while embracing.
Offer my glow throughout the universe and living. Replace it and write from the chart in my notebook or on my special quantum replacings page.

Say out loud my quantumcode with my new and old replacements. I practice seeing new and different movie pictures with each one.

DAY 5
DANCE WITH LIVING

MORNING PRACTICE: Say QC & R.

BATHWAVE TIME FOR LARGE INTESTINE VALUES:	
BATHWAVEs **Part 1: Face & Embrace**	**Quantum Shifts** **Part 2: Replace with Grace**
I face and embrace that I believe nothing changes and I am replacing this.	I can, I will and I enjoy transforming myself.

Replace it **and** write it in my notebook on my special quantum replacings page. Take three deep breaths and dedicate myself to transforming. Focus on pictures of me doing what I truly desire.

EVENING PRACTICE:

BATHWAVE TIME FOR LARGE INTESTINE EMOTIONS:	
BATHWAVEs **Part 1: Face & Embrace**	**Quantum Shifts** **Part 2: Replace with Grace**
I face and embrace that I feel paralysed and immobilised and I am replacing this.	I can, I will and I freely dance with living.

Speak about anything that comes up while embracing. Replace it and write in my notebook on my special quantum replacings page.
Take three deep breaths into belly, chest, and back and see me dancing freely as I desire.

Onward to transfigure my full valuable preciousness of my lungs tomorrow! Good night.

Say out loud my quantumcode with my new and old replacements. I practice seeing new and different moving pictures with each one.

DAY 6:
FAILURE IS NONEXISTENT

MORNING PRACTICE: Say quantumcode and replacements out loud with eyes open.

BATHWAVE TIME FOR LUNG BELIEFS:	
BATHWAVEs **Part 1: Face & Embrace**	**Quantum Shifts** **Part 2: Replace with Grace**
I face and embrace that I am a failure and I am replacing this.	I can, I will and I have everything I need and desire.
I face and embrace that failure exists and I am replacing this.	Everything can, will and is useful. Failure does not exist. I never give up.

Talk about anything I may do that comes up while embracing it. If nothing comes up that's fine. Replace it then writing its correlate from the chart in my notebook on my quantum replacings page. Take three deep breaths and notice how I feel overall and seeing what it looks like living this.

EVENING PRACTICE:

BATHWAVE TIME FOR LUNG ACTIONS:	
BATHWAVEs **Part 1: Face & Embrace**	**Quantum Shifts** **Part 2: Replace with Grace**
I face and embrace that I have to prove myself and I am replacing this.	I can, I will and I am fulfilled as I am.
I face and embrace that I am resisting living fully and I am replacing this.	I can, I will and I embrace living fully.

Talk about anything that comes up while embracing.
Replace it and write the replacement in my notebook on my quantum replacings page.
Take three deep breaths maginating that I am fulfilled completely already. Say out loud my quantumcode with my new and old replacements. I practice seeing new and different movie pictures with each one.

DAY 7:
WONDER OF THE UNIVERSE

MORNING PRACTICE: Say QC & R.

BATHWAVE TIME FOR LUNG THOUGHTS:	
BATHWAVEs **Part 1: Face & Embrace**	**Quantum Shifts** **Part 2: Replace with Grace**
I face and embrace that I reject my values and I am replacing this.	I can, I will and I value myself. I can, I will and I live my values.

Talk about how I may do anything that comes up while embracing it.
Replace it then write it in my notebook on my quantum replacings page. Take three deep breaths and notice how I feel overall.

EVENING PRACTICE:

BATHWAVE TIME FOR LUNG HABITS:	
BATHWAVEs **Part 1: Face & Embrace**	**Quantum Shifts** **Part 2: Replace with Grace**
I face and embrace that nobody can teach me anything and I am replacing this.	I can, I will and I am humble before the beauty and wonder of the universe.

Notice anything that comes up while embracing and replace it. Write replacements from the chart in my notebook on my quantum replacement page. Take three deep breaths and respect myself.
Humble is honest.
Say out loud my quantumcode with my new and old replacements as I desire, visioning each one as I desire...new and different when possible.

DAY 8:
LUNG WORDS AND ATTITUDES

MORNING PRACTICE: Say quantumcode and replacements with eyes open.

BATHWAVE TIME FOR LUNG WORDS:	
BATHWAVEs **Part 1: Face & Embrace**	**Quantum Shifts** **Part 2: Replace with Grace**
I face and embrace that I am the only one who can speak the truth and I am replacing this.	I can, I will and I value the words and needs of others.

While embracing it, notice anything that comes up like guilt or embarrassment. Embrace them if they do. Replace with innocence and self trust and write it to remind myself to live it!
Take three deep breaths visioning myself interacting more easily with others.

EVENING PRACTICE:

BATHWAVE TIME FOR LUNG ATTITUDES:	
BATHWAVEs **Part 1: Face & Embrace**	**Quantum Shifts** **Part 2: Replace with Grace**
I face and embrace that I feel sorry for myself and I am replacing this.	I can, I will and I am enthusiastic about living.

Speak out loud about anything else that comes up while embracing. Replace it and write it in my notebook. Take three deep breaths seeing me enthusiastic living my quantumcode..
Say out loud my quantumcode with my new and old replacements as I desire, visioning each one a reality.

DAY 9:
VALUE AND WORTH

MORNING PRACTICE: Say quantumcode and replacements with eyes open.

BATHWAVE TIME FOR LUNG VALUES:	
BATHWAVEs **Part 1: Face & Embrace**	**Quantum Shifts** **Part 2: Replace with Grace**
I face and embrace that I am responsible for others feelings and needs and I am replacing this.	I can, I will and I am only responsible for me and I honor others to be responsible for their own feelings and needs.

Write about whatever comes up while embracing it. Replace it and write it in my notebook.
I do have responsibilities for my young children, though not totally in control of them. I need to understand they do have sovereignty while being mirror-calls for my creativity. Breathe the replacements in deeply and see if I feel lighter, as if a weight lifted from my shoulders.

EVENING PRACTICE:

BATHWAVE TIME FOR LUNG EMOTIONS:	
BATHWAVEs **Part 1: Face & Embrace**	**Quantum Shifts** **Part 2: Replace with Grace**
I face and embrace that I feel sad and I am replacing this.	I can, I will and I am valuable and worthy.
I face and embrace that I am in grief and guilt.	I can, I will and I am refreshing and energizing

I face and embrace that I feel insecure and unworthy, discouraged, gloomy, hopeless, unhappy, despair and intolerant and I am replacing this.	I can, I will and I am significant. I can, I will and I am living purposeful and enlivening. I can, I will and I am restoring, reviving, & tolerant.

Write about anything that comes up while embracing and replace these. Write replacements in my notebook on my quantum replacement page.
Breathe the replacements in deeply after each one and notice how I feel with all of these. Have a vision of my sacred wholeness living each one.
Say out loud my quantumcode with all my new and old replacements, visioning each one a reality.

DAY 10
TIN MAN'S HEART

MORNING PRACTICE: Say out loud my quantumcode and replacements.
Completion: Practice using my breath today to bring my abidingself matters into focus! Breathe in deeply, taking eternal matters in from around me in that one breath! Then exhale slowly and completely with my focus on concentrating my eternalself to a form I can perceive! This process of concentrating eternal matters to desired forms is the basis of all formed life! This is not a process of manifesting something to me. It is manifesting myself to what I have brought together! Following this process will help me be aware of my illuminatedbeing!

EVENING PRACTICE:
Give my metal Tin Man a heart tonight. My commitment is valuable! To prepare for empowering water add one more BATHWAVE to my growing list.
Am I tempted to sit back and believe that life happens to me? I don't want to do that or I am depowering myself as a victim! Energize my abidingself with this practice today! **I happen to living and since it is recorded in the universe, I am continuing eternally all the ways I live my eternalbeing.**

Repeat this morning's breathing practice.

Say out loud my quantumcode with my new and old replacements. As I speak, I practice watching new and different scenes of my motion picture with me as the star.

QUANTUM REPLACEMENTS FOR MY METAL ORGAN LARGE INTESTINE

I will repeat these present tense replacements for at least 21 days to 3 months (highly recommended). Before bed I say with eyes closed; in the morning I say with eyes open.

I joyfully replace the past to make space for change.
I use my resources benefiting me.

I yield to glow with living.
I manifest sustainable resources.
My words are strong and meaningful.
I keep my promises.
I glow freely through living.
I enjoy transforming myself.
I freely dance with living.

QUANTUM REPLACEMENTS FOR MY METAL ORGAN LUNGS

I have everything I need and desire.
Everything is useful. Failure does not exist. I never give up.
I am fulfilled as I am.
I embrace living fully.
I value myself.
I live my values.
I am humble before the beauty and wonder of the universe.
I value the words and needs of others.
I am enthusiastic about living.
I am only responsible for me* and I honour others to be
responsible for their own feelings and needs.
I am valuable and worthy.
I am refreshing and energizing.
I am significant.
I am living purposeful and enlivening.
I am restoring, reviving, and tolerant.
I happen to living.

DECADE 9

ESSENCE OF WATER IS *NOT* A STATE OF EMERGENCY; WATER IS A STATE OF EMERGENT-SEE MIRACLES

When I am living in a state of emergency, I notice that I am trying to get things to happen. I am attempting to live by causing and effecting.

This is not the case according to the quantum molecular world of water.

Water randomly arises out of the two flammable gases, hydrogen (H2) and oxygen (O2) while not resembling either of them. Water is an emergent miracle!

Is my life an emergent miracle each moment? Yes it is, yet I don't recognize it.

This is how I am doing it:
My constituents and patterns come together as a surprise, an emergent miracle, in each moment. When I don't like the miracle of the moment, all I have to do is add new constituents or patterns that I desire and presto-change-oh!, a new desirable emergent miracle in the present moment.

Mr. Emerchant-Eyes offers water

Water has more, new and different properties than just the two gases.

THIS IS THE SECRET OF WATER

Water is about being able to be aware of my emergent miracles, which means being new and different, replacing the old, in each moment without trying!

As water is new and different from its past (which was a gas!), it is beginning again in each moment!

This is most apparent in the work that I have done with committed couples. One person adds new informotion to their life and the other changes through entinglement, instantly, hence the easy understanding of the process of emergent miracles in relationships.

This realization washes away concern and worry.

What's to worry about when I am able to begin anew in each moment?!

QUALITIES OF WATER

Water amplifies the power of trust, security and confidence.

The essence of water motion is the **gentle power (kidney)** of **self-directing (urinary bladder)** the informotion that I desire to contain or recycle. It's no wonder that the slang American word for urinating is 'pissing' which also means to be expressing one of the forms of anger. When I'm angry my urine turns a darker color of yellow, inviting me to recycle my expression of frustration to one of understanding. I can also embrace my reactivity and replace it with initiating my true desires.

DAY 1
CLARITY HAIKU

Living is clear like
Water: radiant clarity
Is awareness!

Optional last lines:
(Sees perception! or
(Perceives exchange!)

MORNING PRACTICE: Say QC & R. Do dreamwaving if dreams are remembered.
No wonder water is the substance used in rituals signifying new life. Write about things I am going to do today and how I might begin again in each moment. Even if it doesn't seem possible for me to do this, just imagine it and write about it. Keep the awareness of beginning anew throughout my day.

Here are examples by a client:
"Getting ready and going to work: new **"what's to create"** replaces my old **"here we go again."**
Being with the end of intimate relationship: new replacement: **"begin again, what's available, what do I desire?"**, replaces **"the old mental chatter -the urge to submerge in the past!"**
Working day: my emergent idea **"What can I bring and offer?!"** replaces the old **"not this again."**

Sitting in traffic: my emergent thought **"Begin again it's a new moment"** replaces my old **"Are we there yet?"**
Journey across city to men's group after work: emergent miracle thought: **"What do I desire from this, what can I newly create and offer.."** replaces the old **"I should go and I should get it right."**
Hi De hi! Christiansen, T. (2016)

EVENING PRACTICE:
Did I remember to imagine beginning again in any moments of my day? Imagine what my night would be like if I did begin again as I truly desire in my dreams, new and different, in each moment.

Ask for a dream about this! Have no expectations. Record any dream I have in morning.

Read out loud my quantum replacements and eternalself repeating with eyes closed.

DAY 2
WATER CRYSTALLIZES

When water freezes, it crystallizes, which expands it. This gives ice the property of floating, which prevents it from killing all the marine life. This is the only liquid that expands on cooling and therefore floats.

This ability of water to create crystals also enables it to register informotion immediately as in 'entinglement'. Examples of this are shown in research by French scientists, Benveniste, J (1988), Montagnier, L. (2016) and Dr Emoto's research, where water formed different crystalline forms in entanglement with activity, thoughts, or emotions around it Emoto, M. (2005)

Water is demonstrating that it is a molecular vehicle of quantum wave/particle entanglement. All molecules are in entanglement because they are all made of atomic quantum wave/particles, yet not as visibly as water, because all molecules are not as prone to crystallization.

The reason the molecule crystallizes is one of water's most important qualities. It has to do with water's ability to carry the charges of positive (as in protons) on one side and negative (as in the electrons on oxygen) on the other side. This dipole quality of the water molecule gives it the capacity to nurture life.

Oriental medicine calls this ability of water: carrying both yin and yang within itself. Yin and yang are also described as the 'feminine' and 'masculine' qualities. This is an interesting message from water about what nurtures living. WE are mostly made up of water, over 75%. The message of water might look something like I am embracing all aspects of living equally.

MORNING PRACTICE: Be sure to say your quantum wave pattern and replacements after considering any dreams messages.

Imagine I am like a water molecule complete in myself as radiant wavicles reaching the whole universe and crystallizing all my true desires. Picture and write about what I might like to crystallize today in myself that would complete me. Keep it simple and specific. Drink lots of water between meals.

EVENING PRACTICE:
If I don't know Dr. Masaru Emoto's research in *Messages from Water* google it! When in bed talk to my eternalself asking questions before sleep that I would like assistance with in a dream.

Say out loud my quantumcode with my new and old replacements. As I speak, I practice watching new and different scenes of my motion picture with me the star and anyone one else I would like in it.

DAY 3
WATER BELIEVES THERE IS NO SUCH THING AS FAILURE

MORNING PRACTICE:
RECORD any dreams or thoughts/feelings I have as soon as I awake. Say and picture my quantumcode fluidly like the water permeating and BATHWAVEing my cells as well as replacements.

QUANTUM SMILING

All day practice the awareness that I have an abidingbeing... One way is to smile at people as if I see their quantumcodebeings and they are seeing my quantumcode being and I am glad of that. Also, I can smile big because I am glad/proud of myself for knowing my quantum wave patternself. ***See my smile reach the ends of the universe like my quantum wave eternal abidingbeing!***

EVENING PRACTICE:

BATHWAVE TIME FOR WATER BELIEFS:	
BATHWAVEs **Part 1: Face & Embrace**	**Quantum Shifts** **Part 2: Replace with Grace**
I face and embrace that I believe that failure exists, yet I know that it is just a judgment.	I can, I will and I am exploring many ways to do things and never give up.
I face and embrace that I am afraid of my own power, so I don't follow my dreams.	I can, I will and I do appreciate and trust my gentle true strength to follow my dreams.

Add the present tense of each of these two BATHWAVEs to my list of quantum replacements and recite my quantumcode and all replacements!

Day 4
KIDNEY IS GENTLENESS

MORNING PRACTICE: Record any dreams. Say my quantumcode and replacements. I continue my Quantum Smile Practice.
This day Smile with my eternal abidingbeing. Remember my eternal/fraternal twins name and actions and when I smile, I send my essence to the other's illuminatedbeing. My personalbeing is in communion with my abidingbeing as my illuminatedbeing, so that's who is really smiling. Don't worry about feeling or not feeling it; know that it's the vast part of me and the universe gets it, in entinglement!

EVENING PRACTICE:

BATHWAVE TIME FOR WATER ACTIONS:	
BATHWAVEs **Part 1: Face & Embrace**	**Quantum Shifts** **Part 2: Replace with Grace**
I face and embrace that I am not in charge of my living, I am ready to replace this.	I can, I will and I am in charge of my living.
I face and embrace that I am being hard on myself, possibly out of fear.	I can, I will and I am gentle and patient with myself.
I face and embrace that I don't listen to others opinions or needs.	I can, I will and I am open to listening to the opinions and needs of others.
I face and embrace I hold back my energy for fear of ...(anything).	I can, I will and I am freely glowing with energy.

Please add the present tense of each new quantum replacement, to be repeated every morning and night for a few weeks to months. Declare all before bed or as I desire.

DAY 5:
BLADDER IS SELF-DIRECTING

MORNING PRACTICE: After dreamwaving, offer out loud my quantumcode with my new and old replacements. As I speak, I practice watching new and different scenes of my motion picture with me the star.

> **"GIVE YOUR LIFE A DIRECTION TO FOLLOW TODAY!**
> This is not a to-do-list ...it is a direction my life is to take today ...regardless of what may or may not occur! ...write out the direction you desire my life to take today! ...in the past, you've probably set an attitude and followed that at times, ...now, it's time to set an attitude or presence for life to follow! ...you can stop reacting to life and tell my senses what to sense ...and you can also set the character and nature for my life to follow! ...". Cotting, R. B., (2012), Pericope 14A

EVENING PRACTICE: I am doing all these bathwaves and writing the present tense only.

BATHWAVE TIME FOR WATER THOUGHTS:	
BATHWAVEs **Part 1: Face & Embrace**	**Quantum Shifts** **Part 2: Replace with Grace**
I face and embrace that I think I am not free.	I can, I will and I am free in being and doing as I desire.
I face and embrace that I deceive myself and or others and I am ready to replace this as I desire.	I can, I will and I am impeccably honest with myself first and others as I desire.
I face and embrace that I don't believe in myself.	I can, I will and I believe in myself first, and in others as I desire.
I face and embrace that I feel I am controlled by forces or people stronger than my self.	I can, I will and I am SELF-DIRECTING my desires no matter what is happening around me.
I face and embrace that I am indecisive.	I can, I will and I do think clearly and make decisions.

Consider the direction my life seemed to follow today and write a little about what happened?

Before bed, promise out loud my quantumcode with my new and old replacements. As I speak, I practice visioning my new motion picture transforming with me the star.

DAY 6
KIDNEY IS SAFE, SECURE AND CONFIDENT POWER

MORNING PRACTICE: Read my quantum replacements out loud (from the Water and the previous Decades).
Choose a ***new*** direction from my eternalself actions to set my life to follow today.

EVENING PRACTICE: Do all these bathwaves and add to my list.

BATHWAVE TIME FOR WATER HABITS:	
BATHWAVEs **Part 1: Face & Embrace**	**Quantum Shifts** **Part 2: Replace with Grace**
I face and embrace that I am scared to be anything.	I can, I will and I believe I am capable of fulfilling my dreams.
I face and embrace that I belittle myself and diminish my power.	I can, I will and I am uplifting with gentle power in every interaction.

I face and embrace that I have a habit of playing safe and small, because I am afraid to take risks.	I can, I will and I am doing new things.
I face and embrace that I have a habit of believing I am a victim.	I can, I will and I am committed to living a fulfilling life.

Reminder, when I yawn after a quantum replacement, it's a good thing. I am oxygenating new blood which I am manifesting while I am shifting. The informotion of blood is vital emotions.

DAY 7
KIDNEY IS THE CLARITY OF EXPRESSION

MORNING PRACTICE:

Practice directing energy around you to flow through you! ...energy and power ...whether physical or "spiritual" or inspirational do not flow through you naturally! ...and certainly not with intent! ...however, you can induce energy to flow through you ...as music, poetry, healings, perceptions, understanding, or any other form and nature and intensity you desire ...for the intention you desire! Cotting, R. B. (2012), P14a.

EVENING PRACTICE:

What I think are difficulties are like reflections on the surface of water. I embrace difficulties when I dive into it and understand myself and transform my reflection to what I desire.

I am more clear than water. I have the clarity of awareness.

Expanding my clarity with these water words.

BATHWAVE TIME FOR WATER WORDS:	
BATHWAVEs **Part 1: Face & Embrace**	**Quantum Shifts** **Part 2: Replace with Grace**
I face and embrace that I am afraid to speak.	My self-expression can, will and is worth hearing.
I face and embrace that I am shy and timid.	I can, I will and I am safe and confident.
I face and embrace that I have a hard time finding the words I desire.	I can, I will and I am a veritable thesaurus rex, or homo dictionarius. I can, I will and I am expressing myself clearly.

Promise out loud my quantumcode with my new and old replacements.

Note: Practice talking to my eternalbeing out loud, especially when I'm alone. If I have a child who has an imaginary playmate, realise that the child might be speaking to their eternalself. I may ask them questions to learn about its nature and keep notes for their future reference.

Envision the new me out loud.

Day 8
MICROCOSMIC ORBIT TO WATER ATTITUDES OF OPENING AND TRUSTING

MORNING PRACTICE: Say my quantumcode and replacements.

Today I am going to learn a ***simple*** and ***profound*** energy practice called ring around the central nervous system. For yogis it might be called ring around the chakras, or for chi-gong practitioners, the Microcosmic orbit. Chia, M., (2006)

The top of my tongue is resting, touching the roof of my mouth just behind my teeth. Take a deep breath pulling up the muscles of my bottom, and sexual organs (perineum) and imagine fire rising up my spine to the top of my head. Exhaling, picture the fire turning into a waterfall pouring down the front of my body all the way to my sexual organs where it hits the fire, creating steam. When you inhale, pull up the muscles and picture or feel the fire rising up to the top of the head where once again exhale and picture or feel the waterfall pouring down the front of the body again. Repeat the cycle for a total of nine times.

Put one hand over the other, and rub in clockwise circles around my belly button 9 times, Feeling the warmth and the power of the steam that radiates around the body. Relax for one minute, feeling the sensations in my body.

EVENING PRACTICE:
I know what to do...

BATHWAVE TIME FOR WATER ATTITUDES:	
BATHWAVEs **Part 1: Face & Embrace**	**Quantum Shifts** **Part 2: Replace with Grace**
I face and embrace I have to please others and I am ready to replace that attitude.	I can, I will and I do listen and follow my self directing voice.
I face and embrace that I am afraid to live fully.	I can, I will and I do give my whole self to living.
I face and embrace that I distrust people and I'm ready to replace this.	I can, I will and I am discerning and trusting as I desire.
I face and embrace that I close myself to new things and I'm ready to replace this.	I can, I will and I am open.
I face and embrace that living is difficult and a chore and I'm ready for a new option.	Living can, will and is an illuminatedbeing delight.

Promise out loud my quantumcode with my new and old replacements. As I speak, I practice watching new and different scenes of my motion picture with me the star.

DAY 9:
HONEST, GENUINE AND EXPANSIVELY AUTHENTIC

MORNING PRACTICE: Say my QC & R with eyes open.
Dreamwaving and magination practice.

> "**Practice not asking others for permission!** ...you are the authority in my life ...and you have authority, ...therefore, stop seeking approval and permission from others to do what you have the authority to do! ...but, whatever you choose to do, be honest and genuine ...and expansively authentic!" Cotting, R. B. (2018)

EVENING PRACTICE:
New scripting:

BATHWAVE TIME FOR WATER VALUES:	
BATHWAVEs **Part 1: Face & Embrace**	**Quantum Shifts** **Part 2: Replace with Grace**
I face and embrace I value others opinions more than my own and I'm ready to replace that.	I can, I will and I do value my self-directing voice.
I face and embrace that I take better care of others than I take care of myself and I'm ready to replace that with the airplane principle. *Remember on an airplane when an oxygen mask drops down and the stewardess says please place on my own mask before you attempt to assist anyone else.*	I can, I will and I do take care of myself first, as well as taking care of others.
I face and embrace that I am living for my past or my future and I'm ready to replace that.	I can, I will and I transform my past to my true desires in this present moment.

Say out loud my quantumcode with my new and old replacements as if I am auditioning for a movie starring with my favorite actor/actress. Picture getting my part.

DAY 10
MELTING WATER'S ICEBERG OF FEAR

MORNING PRACTICE: I say my quantumcode and replacements.

DREAMWAVE AND ORGANIZE ALL YOUR NEW QUANTUM REPLACEMENTS

OVERVIEW:

I keep facing my fear as I am doing. The warmth of the act of embracing with no judging melts it. For most of my fear is an iceberg. The good news is ICE FLOATS! So as I embrace what is showing it rises up from down under until eventually the iceberg is gone. This, of course, is quickened in the quantum wave of living. All the water that was frozen in me now supports ME living fully!!

That is one of the reasons water is the basis of material life.

Material life is part of quantum wave living.

The quantum wave is the invisible, yet physical part of quantum wave living.

As I live these two identities of me I have the third identity called my new 'wavicle', (my wav/icle=**wave**/part-**icle**). The new wavicle is my new particle body communicating awarefully with my quantum wave partner. The **wavicle** is my **resplendent illuminatedbeing which is universalself. My particle with quantum wave wings to soar and score my authentic being. Hence I call myselves a triunity (trinity); there are three selves of me.**

I know what to do...

BATHWAVE TIME FOR WATER EMOTIONS:	
BATHWAVEs **Part 1: Face & Embrace**	**Quantum Shifts** **Part 2: Replace with Grace**
I face and embrace that am scared, fearful and afraid, and I am ready to replace the whole shebang.	I can, I will and I am confident and secure, even with the unknown.
I face and embrace that I am cold-hearted and I'm ready to warm things up.	I can, I will and I am generous and warm-hearted.
I face and embrace I doubt myself and others I am ready to expand that.	I can, I will and I do believe in myself and others as I desire.

ADD TO MY Quantum Replacements.

EVENING PRACTICE:

BATHWAVE TIME FOR WATER EMOTIONS:	
BATHWAVEs **Part 1: Face & Embrace**	**Quantum Shifts** **Part 2: Replace with Grace**
I face and embrace I am jealous or envious of myself or others. This means I do not believe I am as brilliant as OTHERS and I'm ready to replace that.	I can, I will and I am wonderful just the way I am.
I face and embrace that am scared, fearful and afraid, and I am ready to replace the whole shebang.	I can, I will and I am confident and secure, even with the unknown.
I face and embrace that I am cold-hearted and I'm ready to warm things up.	I can, I will and I am generous and warm-hearted.

Say out loud my quantumcode with my new and old replacements. As I speak, I practice watching new and different scenes of my new motion picture STARRING ME, ALONG WITH ANOTHER FAVORITE STAR OR PERSON.

QUANTUM REPLACEMENTS FOR MY WATER ELEMENT KIDNEYS & BLADDER

I will repeat these present tense replacements for at least 21 days to 3 months (highly recommended). Before bed say with eyes closed. In the morning say with eyes open.

I am exploring many ways to do things and never give up.
I do appreciate and trust my gentle true strength to follow my dreams.
I am in charge of my living.
I am gentle and patient with myself.
I am open to listening to the opinions and needs of others.
I am freely glowing with energy.
I am free in being and doing as I desire.
I am impeccably honest with myself first and others as I desire.
I believe in myself first, and in others as I desire.
I am SELF-DIRECTING my desires no matter what is happening around me.
I do think clearly in making decisions.
I believe I am capable of fulfilling my dreams.
I am uplifting with gentle power in every interaction.
I am doing new things.
I am committed to living a fulfilling life.
My self-expression can, will and is worth hearing.
I am safe and confident.
I am a veritable thesaurus rex, or homo dictionarius.

I can, I will and I am expressing myself clearly.
I do listen and follow my self directing voice.
I do give my whole self to living.
I am discerning and trusting as I desire.
I am open.
Living can, will and is an illuminatedbeing delight.
I do value my self-directing voice.
I do take care of myself first, as well as taking care of others.
I transform my past to my true desires in this present moment.
I am confident and secure, even with the unknown.
I am generous and warm-hearted.
I do believe in my self and others as I desire.
I am wonderful just the way I am.
I am confident and secure, even with the unknown.
I am generous and warm-hearted.

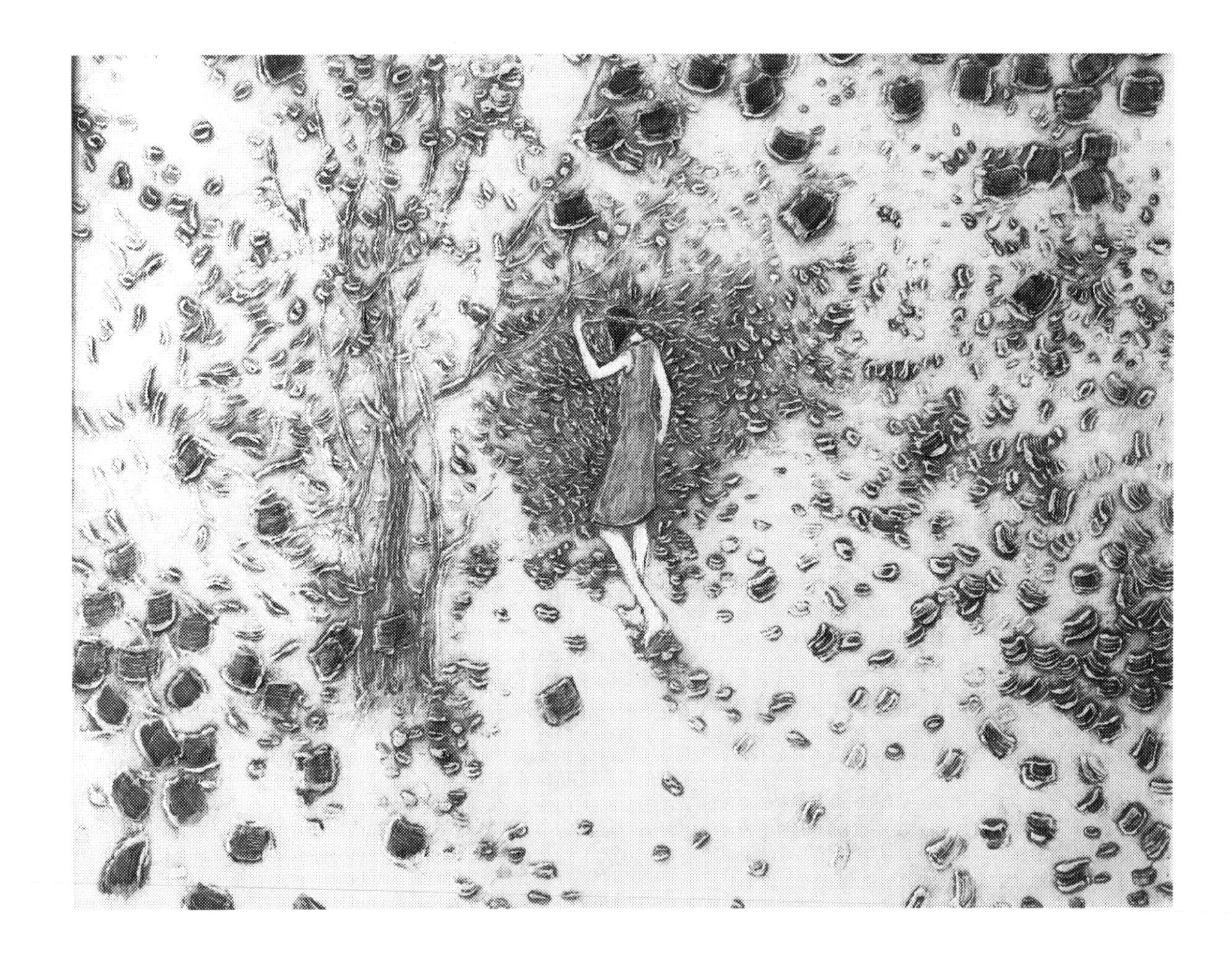

DECADE 10

WOOD MOTION RELATING: WOOD I OR WOODN'T I BEE TREE-MEND-US.

THROUGH ENTINGLEMENT HAIKU

My transformation
is my invitation to
Have the world my wave!

WOOD IS THE SEED OF SELF-INITIATION

Wood begins as a seed of random **self-organizing** data and informotion of my recurring eternalbeing.

When the seed of my resplendency **sprouts**, the sunlight hits the new leaves in a cascade of **entanglement**. All the molecules of photosynthesis **IN INSTANT synchrony**, efficiently and effectively send **energy down my roots to reach deep into my earthen personalbeing**. I am fertilized by the **constituents and compost I amass**. I water myself with my roots absorbing from emergent fluidity of the vital e-motions I desire.

The wind blows air's guidance of self worth, **pollinating** my personalbeing with my eternalbeing's **pulsating patterns**, enabling the ordinary extraordinary **fruition** of my ordinary resplendent actions of sharing myself.

My roots are absorbing the emergent waters of my miraculous life. Freely flowing fluid rises, just like in my spine, to combine with the CO2 my leaves inhale, **manifesting** the next emergent miracle as the sweetness of fruits, flowers, and food, which I **offer** to all who ask.

As I **outgrow** my be-leafs, the guidance of metallic shearing called pruning of my old overlapping branches of habits, attitudes and values, enables **replacement** with new radiant expansion.

The **blossoming and ripening** of my sweet resplendency is evident as I **reach** up with my new branches into the fire of my eternal passion to become a burning bush!

SUMMARY OF WOOD PROCESSES IN ME:

1) Germinate and initiate self-organizing seed = believe

2) Sprout= act

3) Entinglement= instant synchrony= root deep /branch out in universe= thinking

4) Pollinating is Beeing Pulsating patterns= habiting

5) Blossoming = Fruiting and ripening = wording

6) Outgrowing into my "environment" of sun and water vapor= attituding

7) Replace old past patterns with new Atom and Weave (we've)= valuing

8) Reaching out through universe that is raining new informotion= radiant emoting

DAY 1
GERMINATE AND INITIATE SELF-ORGANIZING SEED = BELIEVE

MORNING PRACTICE: Say my quantumcode and replacements with eyes open. Maginate creating new seeds of actions all day long. I am planting those seeds which grow my abidingself (AS) shape and form. Maginate what kind of seeds my quantumcode self would like to plant this day? Jot down AS actions in list form in my notebook.

Write words as seeds on a card or my hand to remind me.

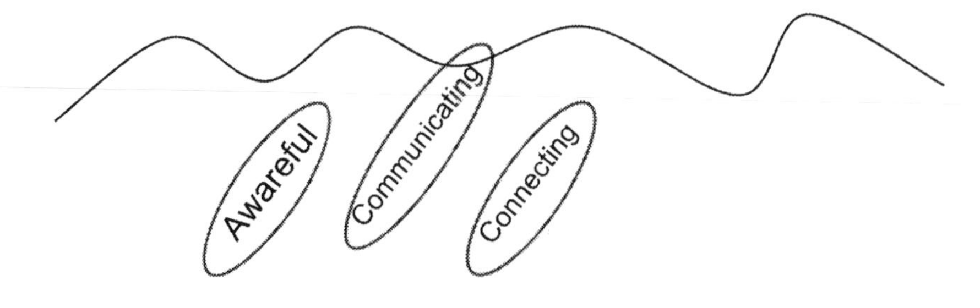

EVENING PRACTICE:
Did I maginate (vision) any actions as seeds today? I note a few down in my notebook. Are there any that come to me now? Did any new actions result from these seeds? Ask for a seed dream. Record on waking and dreamwave it!

Speak and envision my quantumcode and replacements of the last 3 Decades or more, out loud eyes closed.

DAY 2
REPLACING OLD WOODY BELEAFING

MORNING PRACTICE: Speak and envision my quantumcode and replacements of the last 3 Decades or more out loud eyes open.

> "**You already are an eternalbeing!**...so practice feeling myself as an eternalbeing...just as easily and completely as you feel myself as a personalbeing! ...for all difficult and limiting problems and afflictions of being a personalbeing can be better-understood and dealt with in the awareness and context of my being eternal!
>
> Direct perception practices and dreams enable you to stand directly in the shoes of an eternalbeing...which provides a different perspective of everything in my personal life ...should you desire such a perspective!
>
> There is no struggle or discontent in my body or life except in my mind! ...therefore, when my mind is fully active and not discontent ...there is no conflict ...and my eternalself can become a loving presence in my body and life for you to bring alive and personalize! ...for my eternalself does not act through you, you act for my eternalself! ...or you act as an eternalbeing with all the power and awareness of an eternalbeing! ...but the choice to live as a fully empowered personalbeing·eternalbeing is completely up to you!
>
> There is no finer expression of life than being alive and not discontent!" Cotting, R. B., (2012), p. 14a #15

Spend my day stopping any discontenting or upsetting of myself no matter what goes on. It's even possible to be angry without discontenting myself if I need that. Keep track of how many times I discontent myself.

EVENING PRACTICE:

BATHWAVE TIME FOR WOOD BELIEFS:	
BATHWAVEs **Part 1: Face & Embrace**	**Quantum Shifts** **Part 2: Replace with Grace**
I face and embrace that I block and sabotage myself.	I can, I will and I do what benefits me.

Vision myself as my favorite tree with leaves shining with sunlight and moonlight or needles all pining for illumination, lol. Add the present tense to my list and say them all with my quantumcode.

DAY 3:
SPROUTING= ACTION

MORNING PRACTICE: Envision while I declare my promises.

BATHWAVE TIME FOR WOOD ESSENCE	
BATHWAVEs **Part 1: Face & Embrace**	**Quantum Shifts** **Part 2: Replace with Grace**
Face and embrace that I am discontenting myself and I would like to replace that behavior.	I can, I will and I do stop discontenting myself today.

Be aware of myself as much as possible today replacing upsets again.

EVENING PRACTICE:

BATHWAVE TIME FOR WOOD ACTIONS:	
BATHWAVEs **Part 1: Face & Embrace**	**Quantum Shifts** **Part 2: Replace with Grace**
I face and embrace that I react.	I can, I will and I do listen and initiate what I truly desire.
I face and embrace that I put off deciding until it's too late.	I can, I will and I make timely decisions easily.

DAY 4
THOUGHTS OF WOOD DOINGS

Entinglement = instantly synchronous = rooting deeply = universal deep thinking

MORNING PRACTICE: Say my quantumcode and replacements with my eyes open.

> **"Pay close attention to my attitudes and emotions!** ...don't just sense them and feel them ...and then dismiss them or struggle with them, for they are not just reactions to my experiences and judgements, they actually limit and control you ...and frequently create the events and conditions you are experiencing and judging, and they greatly influence my life and body ...and my eternalself! ...actually, my attitudes and emotions carry my discontent into every situation ...which keeps you endlessly reacting ...and further discontenting my attitudes, emotions, and body ...rather than being self-evidently communioned with my eternalself as you truly desire!" Cotting, R. B., (2012), P14a #35

Be aware of when I react and judge, before I upset myself!

Ask myself what I would rather initiate, to the tune of my eternalself, and unconditionally live and love myself. In others words stop judging myself and begin again.

EVENING PRACTICE:

BATHWAVE TIME FOR WOOD THOUGHTS:	
BATHWAVEs **Part 1: Face & Embrace**	**Quantum Shifts** **Part 2: Replace with Grace**
I face and embrace that I am resenting others or myself.	I can, I will and I do recognize that everything is beneficial.
I face and embrace I am disappointed in others or myself.	I can, I will and I do recognize the eternalbeing in others and myself.
I face and embrace I have trouble making choices I desire.	I can, I will and I do live my true desires with ease.

DAY 5
POLLINATE HABITS I DESIRE

Pollinating is Pulsating Patterns = Habits

MORNING PRACTICE: After dreamwaving, please reveal my quantumcode and replacements like it is the first time ever.

Reread and repeat yesterday's morning practice. See if I can catch myself before I react to discontent and then just smile to myself and begin again.

EVENING PRACTICE:

BATHWAVE TIME FOR WOOD HABITS:	
BATHWAVEs **Part 1: Face & Embrace**	**Quantum Shifts** **Part 2: Replace with Grace**
I face and embrace I am staying as I am.	I can, I will and I am pulsating my transformation throughout the universe.
I face and embrace I am addicted and stuck to throwing in my towel to a dramatic pity party.	I can, I will and I am throwing off the towel like the veil at my belly dancing party.

When I read and recite my abidingself and replacements offer them as if the existence of the universe depends on it and it does!

DAY 6
BLOSSOM AND RIPEN MY WORDS

Fruiting = Blossoming and ripening = words

MORNING PRACTICE: Recite my abidingself and replacements with eyes open.
OFFERING TO MYSELF SWEET WORDS

> **"The greatest meaning of life is not that of serving others …it is offering to life! …for in that you offer to myself!** …do everything as an offering to life …and my life and understandings will change substantially …and immediately! …but only if there are no special or compassionate offerings, no right or divine offerings, and no actions of service …all actions are my offerings! …and do not treat communioning with my eternalself as anything but my offering to the world around you!
>
> An expansive and extensive expression of myself as an eternalbeing requires a new and different agreement with the world around you! …in the past, my agreement was that the universe would be to you as you are to the universe! …and from that, you felt you would be rewarded for right thinking and good deeds! …that you would be loved as you love others …that it will be done unto you as you do unto others! …and you would be served as you serve others! …all of this was unfortunately expressed as, "Do unto others as you would have others do unto you"!
>
> Certainly, you are now willing to change my agreement with life …such that the more you offer …the more you will discover you have to offer! …the more you communion …the more you will be communioned! …that service to others is actually service to myself! …and you are expanded only by what you communion with …such

that you have my eternalself and life only as you give life to my eternalself …and communion that with my personalself and life!

Every act of life is complete in itself! …there is no reward or punishment! …no return for actions! …and no quid pro quo! …for what you communion is what you are …and what you live and have! …you do not reap what you sow …you live what you sow! …in fact, though you don't seem to realize it, you are what you sow! …and every offering is a completed action!" Cotting, R. B., (2012), p. 14a#13

Focus on offering my quantumwave pattern to live today! Designate every 3 hours to check in and recite my authentic pattern, say, at these hours: 12, 3, 6, 9...repeat, or just when I remember. Be gentle with myself!

EVENING PRACTICE:

BATHWAVE TIME FOR WOOD WORDS:	
BATHWAVEs **Part 1: Face & Embrace**	**Quantum Shifts** **Part 2: Replace with Grace**
I face and embrace I use abrupt angry words.	I can, I will and I do hear and empathise with other people's feelings and needs.
I face and embrace I limit myself to brain thinking.	I can, I will and I do think with the universe thanks to entinglement.
I face and embrace I have a hard time finding words.	I can, I will and I do find words with the motion of my body and the entinglement of the universe.

Do my evening revelations.

DAY 7
EXPANDING ATTITUDES OF THE ENVIRONMENT I DESIRE

Outgrowing my environment = attitudes

MORNING PRACTICE: Say my abidingself and replacements.

"Decide and declare, if only to myself, that you are meeting and experiencing my eternalself in every event and action! …begin every situation and every moment of my life with a firm declaration of who you are …and what you are doing! …not what you want to do or will do …or what you intend to do! …and certainly not who you hope to become! …and do not begin with an "open mind" …that you accept what comes to you! …for everything in my life depends upon what you declare and commit to …not on what you accept or allow!

> Do not struggle ...or stress myself to fulfill my decisions and declarations, ...remain assured ...and do not discontent myself! ...and carry my promises and commitments with you in my daily affairs ...in everything you do!" Cotting, R. B. (2012)

Reread yesterday's whole quotation and see if it makes more sense. If not have a friend read it and discuss it together.

Be open to offering my eternal self today in some new way to life.
If nothing comes to my mind, I can do the Asian smile with a bow of the head in reverence when I see anyone. Think that I am bowing to their eternal self. When I can give the bow a slight flavor of my eternal self I am beginning my offering.

EVENING PRACTICE:

BATHWAVE TIME FOR WOOD ATTITUDES:	
BATHWAVEs **Part 1: Face & Embrace**	**Quantum Shifts** **Part 2: Replace with Grace**
I face and embrace I am depleting my vital emotions.	I can, I will and I am filling myself with emotions of my heart's desire.
I face and embrace I am an irritated and or irritating.	I can, I will and I do breathe gratitude into every moment.
I face and embrace I am bored.	I can, I will and I do stimulate my own opalescent (sparkling) radiance.

DAY 8
VALUE AND MEANING OF MY UNIQUE QUANTUMCODE IS MY OFFERING

Replacing old values = abidingself values

MORNING PRACTICE: Recite my quantumcode and replacements this morning.

> *"Practice not planning your life!* ...assuming you understand that ensuing moments and events replace prior moments and events, then you understand that neither moments or events are sequential! ...yet, you plan your life sequentially according to language! ...this is why life seldom turns out according to your plans! ...you and life are not on the same track! ...so, why plan? ...instead, create options ....give yourself options! ...which are far-more expansive and "freeing" than alternatives!
>
> Be careful not to create or plan options in an order of desirability ...that's planning again! ...and, in that, if you are not following plan A, your life is less than desirable! ...options should always be other ...or to take the place of ...never

"better or lesser"! ...and do not consider options as preferred ...or not-preferred, but acceptable! ...options as "other" always gives you freedom!

With options, there are never optimal moments or conditions! ...never any failures or losses! ...never any disappointments! ...for every option is sufficient and complete in its own right ...or is easily and comfortably replaced by another option!

Options have no goals, conclusion, or endings! ...for those destroy options!"
Cotting, R. B., (2012) P. 14a #39

To find options, I ask for a dream, or see the mirror-call to replace in the aspect of myself that it mirrors. Be gentle with myself!

EVENING PRACTICE:

BATHWAVE TIME FOR WOOD VALUES:	
BATHWAVEs **Part 1: Face & Embrace**	**Quantum Shifts** **Part 2: Replace with Grace**
I face and embrace I use am using guilt to get things done.	I can, I will and I do have the capacity and willingness to be flexible – to fearlessly leap from each sub quark of the time-continuum to the next.
I face and embrace I think I am guilty.	I can, I will and I am innocent, refreshing and free of coercing (forcing).

DAY 9
WOOD I LIKE TO BE THE REASON FOR MY ILLUMINATEDBEING

QUANTUM WAVE HAIKU

As quantum wave I
Don't live feelings. I live
My abidingbeing!

MORNING PRACTICE: Say my quantumcode and replacements. When completed, repeat yesterday's morning practice during the whole day.

The greatest meaning of life is not that of serving others ...it is offering to life! ...for in that I offer to myself!

EVENING PRACTICE:

Reaching out through universe = radiant emotions

BATHWAVE TIME FOR WOOD EMOTIONS:	
BATHWAVEs **Part 1: Face & Embrace**	**Quantum Shifts** **Part 2: Replace with Grace**
I face and embrace I am angry.	I can, I will and I am refreshing and soothingly buoyant.
I face and embrace I am enraging.	I can, I will and I relax and do as I desire.
I face and embrace I am burning myself up.	I can, I will and I do fill myself with emotions of my hearts desire.
I face and embrace that I am frustrated.	I can, I will and I do begin again new and different in each moment.
I face and embrace that I discontent myself.	I can, I will and I do I stop discontenting myself. I can, I will and I do I live with my eternalself in one-nest.

DAY 10
HOT WOODY EMOTIONS:
AM I DOING TREE-MEND-US?

MORNING PRACTICE:

"Practice being content …even while upset or angry! …being content does not mean being satisfied or fulfilled …it means not discontenting yourself! …being angry or upset certainly seems to indicate you are not satisfied or fulfilled …but you don't have to be discontent about that! …your head says "Do this!" and your heart says, "Do that!" …and your mind reminds you of your promises, dedications, and inspirations! …will this create a conflict for you? …is there anything you can do that won't annoy you …and discontent you in some way, no matter what you decide?

You frequently have mixed emotions about the same thing! …and you often struggle with mixed emotions …such as being both pleased with and angry at something …or a person or an event at the same time! …this disagreement between "head and heart" …or between conflicting emotions and mind is a constant struggle in your life …why is that? …what's going on? …do you know? …is it a normal and natural doubt and uncertainty? …or is something else happening?

Are you both pleased and angry? …is that possible? …are you alternately pleased then angry then pleased? …are you pleased about your anger? …or angry at your pleasure? …are you simultaneously in control and out of control? …or is it possible that one aspect of you is angry …and another part is pleased? Cotting, R. B., (2012) P.14a #10

EVENING PRACTICE:
Complete woods hot emotions: (and prevent a forest fire)

BATHWAVE TIME FOR WOOD HOT EMOTIONS:	
BATHWAVEs **Part 1: Face & Embrace**	**Quantum Shifts** **Part 2: Replace with Grace**
I face and embrace I am resenting.	I can, I will and I am appreciating my mirror–calls.
I face and embrace I am bored and neglected.	I can, I will and I am creative and engaging.
I face and embrace I am indecisive and procrastinating.	I can, I will and I am excited, decisive and just doing it.
I face and embrace that I am depressed and nervous.	I can, I will and I am patient and self motivating in each moment.
I face and embrace that I am disappointing/ed and self sabotaging.	I can, I will and I am cooperating with myself without judging. I can, I will and I do live as my universal illuminatedbeing authentically.

AM I feeling tree-mend-us?

Sorry, I've been a punster a longo time and quantum wave living hasn't replaced that. It's actually getting stronger?!

As you see, I am not talking about planting trees here. Although, those attracted to doing this have practiced a process of visioning and acting on it. This is what I am talking about. Initiation is in each moment. Initiation is what every mystery school or club has ever encouraged me to do. I believe the quantum world says,

I CAN ONLY INITIATE MYSELF (MYSELVES).

Any resemblance to planting trees is purely intentional to show me how easy this process is. EXCEPT IN QUANTUM WAVE LIVING IT ALL HAPPENS AT ONCE: NO SEQUENTIAL PROCESS AS IN THE GROWING OF THE TREE.

ALL THREE OF ME AS VIRTUAL SELVES AND ALL THREE OF ME AS BEINGS ARE ALL HERE IN THIS MOMENT RADIANT/ RESPLENDENT. See figure of six-pointed star that I am.

Wood BATHWAVEs are only a bridge, which I do not need once I am direct perceiving or visioning living as my eternalbeing.

Write about what that looks like to me now that I have completed what I wood do yet I wood not?!

And listen and speak all my quantum replacements along with eternalself before sleep.

QUANTUM REPLACEMENTS FOR MY WOOD ELEMENT LIVER & GALLBLADDER ORGANS

I will repeat these present tense replacements for at least 21 days to 3 months (highly recommended). Before bed say with eyes closed. In the morning say with eyes open.

I do what benefits me.
I do stop discontenting myself today.
I do listen and initiate what I truly desire.
I make timely decisions easily.
I do recognize that everything is beneficial.
I do recognize the eternalbeing in others and myself.
I do live my true desires with ease.
I am pulsating my transformation throughout the universe.
I am throwing off the towel like the veil at my belly dancing party.
I do hear and empathise with other people's feelings and needs.
I do think with the universe thanks to entinglement.
I do find words with the motion of my body and the entinglement of the universe.
I am filling myself with emotions of my heart's desire.
I do breathe gratitude into every moment.
I do stimulate my own opalescent (sparkling) radiance.
I do have the capacity and willingness to be flexible – to fearlessly leap
from each sub quark of the time-continuum to the next.
I am innocent, refreshing and free of coercing. (forcing)
I am refreshing and soothingly buoyant.
I relax and do as I desire.
I do fill myself with emotions of my hearts desire.
I do begin again new and different in each moment.
I do I stop discontenting myself.
I can, I will and I do I live with my eternalself in one-nest.

QUANTUM REPLACEMENTS FOR MY HOT EMOTIONS:

I am appreciating my mirror–calls.
I am creative and engaging.
I am excited, decisive and just doing it.
I am patient and self-motivating in each moment.
I do live as my universal illuminatedbeing authentically.

QUANTUM WAY OF LIVING
HOW YOU MANIFEST THE UNIVERSE

DECADE 11

HEART FIRE MOTION

FIRED UP RADIANT MOTION:
ESSENCE OF YOUR FIRE IS REACHING
AS OUTGOING QUANTUM WAVE

What Does Quantum Radiance Look like in Living?

The four main motions of fire are described in these actions:

IGNITE
ACTIVATE
EVAPORATE
CONCENTRATE
ACCUMULATE
PRECIPITATE
REACH OUT
ORIGINATE
INITIATE
SELF-NURTURE
ASSIMILATE
PARTICIPATE
RADIATE

Radiance is the action of passion that ignites and evaporates cold wet BATHWAVEs like insecurity, contraction, numbness, grief and fear into steam. The steam then rains down IN WARMTH without judging, replacing coldness and reaching out as:
authenticity, (heart) IGNITES bonding, (pericardium = the waterbed of the heart) CONCENTRATES appreciation of the constant existence of the "harmony of entinglement" (triple heater = circulation of energy), REACHES OUT and absorbing nourishment from this (small intestine) SELF-NURTURES.

The informotion of these replacements reaches out equally in all directions to the universe.
That is the beginning of radiance, which is the 'resplendance' of communion with our selves.
Now that we are resplendent with the communion of our personalbeing and eternalbeing the radiance expands. Our resplendentbeing enables us to be in 'entinglement' with anything around us as conditions of expansion (COE) and then we "particularize" that as we desire. Radiance is this physical reaching forth thru entinglements in resplendency (without cause and effect) to conditions of expansion.

Intimacy becomes the entinglement of two resplendentbeings. Their emergent mirror-calls are the expansion of the universe.

The fire element is sparked with the wonderful realization that everything is a mirror-call of our resplendency. This is the dance of relationship, the dance of resplendency. Since everything in our living is a relationship, this is the dance of the expansion of our universe as I live my authentic abiding being.

DAY 1
WARMING THE FIRES OF MY HEART

MORNING PRACTICE: Heart Motion
Recite my quantumcoade and replacements. Starting this moment and continuing throughout my day, without judging myself, notice when my self closes down or chooses coldness, looking for a stimulant to warm up (like HOT coffee, tea, chocolate, most kinds of pepper or spices or a hug). Does this coldness prevent me from bonding my whole being in relationship to myself or others? Observe my BATHWAVEs today and notice how I would like to warm up my bathwaves. Jot down or audio record my thoughts of my unmet needs for warmth, connection and bonding and the feelings that arise from those unmet needs.

EVENING PRACTICE:
Write the unmet needs I observed today in myself. Jot down some of the BATHWAVEs I would like to add to warm myself up?
Entinglement: Picture myself living this way regardless of others attitudes or responses. Imagine extending warmth, connection, and bonding to everyone tomorrow with a smile and more when possible.
Before I sleep say my quantumcode and replacements.

DAY 2
BATHWAVE TIME FOR HEART BELIEFS:

MORNING PRACTICE: Say or sing my quantumcode and replacements.

> Practice believing you have an eternalself! ...you say you believe you have an eternalself ...and maybe you do! ...but, if you are like most people, you need a little practice believing, ...simply because saying you believe you have an eternalself is not what you really believe until you are awarefully that belief! ...for, believe it or not, stripped of its distortions *believe* means to give credence to! ...therefore, if you haven't openly and awarefully expressed myself as an eternalbeing, and unless you are unquestionably living as a communioned personalbeing·eternalbeing, you cannot claim to be giving credence to an eternalbeing! ...ohhh, certainly, you may think this is so or hope it is so, but you don't really believe it!

> Isn't it reasonable to declare that if you don't live a belief ...you don't really believe it? ...besides, you only have the right to declare what you are truly living! Cotting, R. B., (2018).

Pick 3 hours (like noon, 3PM and 6PM) of my day to stop and ask myself what percent of the hour did I live my quantumcode in hindsight?

See if this helps me to live it?

EVENING PRACTICE:

BATHWAVE TIME FOR HEART BELIEFS:	
BATHWAVEs **Part 1: Face & Embrace**	**Quantum Shifts** **Part 2: Replace with Grace**
I face and embrace my love letter of wanting to be perfect, and believing that perfect exists.	I can, I will and I live my eternal authentic and real self.
I face and embrace my love letter that bad things happen.	I can, I will and I know everything is a sacred mirror doorway of my whole self.
I face and embrace my love letter that I'm a stranger to myself.	I can, I will and I am replacing my love letter with the miracle of being aware of my whole self.
I face and embrace my love letter that I am judging myself harshly.	I can, I will and I am replacing my love letter with I love my aliveness unconditionally.

Write down the present tense only of Part 2s in my notebook, or not, since there is a list at end of chapter. Say all of my replacements and quantumcode before I sleep.

DAY 3:
WARMING UP FOR FOR HEART ACTIONS:

MORNING PRACTICE: Dreamwaving, Quantumcode and Replacements

> **"Practice being everything you do!** ...like it or not, language only has a place for you as the causer of things ...or the one to whom things happen! ...while life also has a place and life for you as the action itself ...and the issue of myself as an action! ...however, you have forsaken the expansive and extensive nature of life for the false comfort and security of language! ...does it seem wise to continue in this primitive choice and way of life? ...are you interested in being and experiencing all you are?" Cotting, R. B., (2012), #20 in Pericope 14a

Keep a record of the things I do today to recall this evening when I transform my heart actions!

EVENING PRACTICE:

BATHWAVE TIME FOR HEART ACTIONS:	
BATHWAVEs **Part 1: Face & Embrace**	**Quantum Shifts** **Part 2: Replace with Grace**
I face and embrace my love letter: I hide from myself.	I can, I will and I am replacing my love letter with the miracle of awakening my whole self.
I face and embrace my love letter that I am shy or frightened in some way or another.	I can, I will and I am safe and secure no matter what's happening.
I face and embrace my love letter that I deny myself, which creates my unconsciousness.	I can, I will and I am replacing my love letter with the miracle of appreciating how alive I am.
I face and embrace my love letter of wanting to be perfect, that I withhold my gifts of friendship.	I can, I will and I am replacing my love letter with I listen to, acknowledge and communicate the feelings and needs of my friends.

Write down the present tense only of Part 2s in my list then state quantumcode and replacements before I sleep.

DAY 4:
THINKING WITH MY HEART'S UNIVERSE

MORNING PRACTICE: Say my quantumcode and replacements this morning with my eyes open.

> **"You already are an eternalbeing! ...so practice feeling myself as an eternalbeing ...just as easily and completely as you feel myself as a personalbeing!** ...for all difficult and limiting problems and afflictions of being a personalbeing can be better-understood and dealt with in the awareness and context of my being eternal!
>
> Suscitation* practices and dreams enable you to stand directly in the shoes of an eternalbeing ...which provides a different perspective of everything in my personal life ...should you desire such a perspective!
>
> There is no struggle or discontent in my body or life except in my mind! ...therefore, when my mind is fully active and not discontent ...there is no conflict ...and my eternalself can become a loving presence in my body and life for you to bring alive and personalize! ...for my eternalself does not act through you, you act for my eternalself! ...or you act as an eternalbeing with all the power and awareness of an eternalbeing! ...but the choice to live as a fully empowered personalbeing·eternalbeing is completely up to you!
>
> There is no finer expression of life than being alive and not discontent! Cotting, R. B., (2012) Pericope14a

Create a reminder for myself that I have a eternal companion with me that is physical yet invisible named________________________________.

EVENING PRACTICE:

BATHWAVE TIME FOR HEART THOUGHTS:	
BATHWAVEs **Part 1: Face & Embrace**	**Quantum Shifts** **Part 2: Replace with Grace**
I face and embrace my love letter: that I belittle myself.	I can, I will and I am replacing my love letter with having uplifting, complimentary thoughts of myself and then others.
I face and embrace my love letter that I am avoiding thinking about myself.	I can, I will and I am replacing my love letter with I am open and aware of myself.

Write down the present tense only of Part 2s in my list and recite all including my quantumcode.

DAY 5
HEART HABITS

MORNING PRACTICE: Say my quantumcode and replacements this morning with my eyes open.

> "Begin observing and feeling myself! ...feel my personalself communioned with my body! ...listen carefully to what you are truly saying ...not merely to the words! ...and practice stepping away from whatever you are doing to perceive myself! ...don't just watch what you are doing!
>
> You've probably felt sick ...and even felt pleased and happy, ...but have you ever felt myself feeling sick or pleased or happy? ...and if you think carefully about it, can't you observe myself being sick or pleased ...or happy?
>
> Practice feeling myself! ...not just feeling my body ...or feeling the same as my body feels! ...feel the self inside my body! ...then practice watching that self within my body!
>
> When you hurt myself, did you do that to myself ...or to my body? ...and who saw you do that? ...are you willing to give my full attention to everything you are doing and saying at every moment? ...and simultaneously pay close attention to you ...saying and doing? ...I am my own witness! ...what would you like to see myself doing ...and saying?" Cotting, R. B., (2012) P. 14a

So watch myself with no judging, being, saying and doing what I truly desire as much as possible! Keep notes about BATHWAVEs you wish to do tonight!

EVENING PRACTICE:

BATHWAVE TIME FOR HEART HABITS	
BATHWAVEs **Part 1: Face & Embrace**	**Quantum Shifts** **Part 2: Replace with Grace**
I face and embrace my love letter that I run away from relationships and myself.	I can, I will and I am replacing my love letter with the miracle of bringing myself anew to each relationship and myself in each moment.
I face and embrace my love letter that I isolate myself as a loner.	I can, I will and I am replacing my love letter with the miracle that I reaching out to communicate in relationship.
I face and embrace my love letter that I deny my entinglement mirror-calls.	I can, I will and I am replacing my love letter with appreciating the power of my entinglement.

Write down the present tense only of Part 2s in my notebook (or use list at end of this Decade) and say all replacements and quantumcode out loud before sleeping.

DAY 6:
WARM HEARTENING WORDS

MORNING PRACTICE: DREAMWAVE (DW), QUANTUMCODE (QC) AND REPLACEMENT REMINDING

Reread and follow directions from yesterday's morning practice. Do again!

EVENING PRACTICE:

BATHWAVE TIME FOR HEART WORDS:	
BATHWAVEs **Part 1: Face & Embrace**	**Quantum Shifts** **Part 2: Replace with Grace**
I face and embrace my love letter that I stifle and judge everything I say.	I can, I will and I am replacing my love letter with I support resplendence with my communication.
I face and embrace my love letter that I find it hard to talk to people.	I can, I will and I am replacing my love letter connecting and conversing easily with people.
I face and embrace my love letter that I can't find the words to say what I desire.	I can, I will and I am replacing my love letter with I speak my feelings and needs honestly to others.

Write down the present tense only of Part 2s in my notebook or use end of decade. Say QC & R.

DAY 7:
IGNITE MY ATTITUDE OF GRATITUDE

MORNING PRACTICE: DW AND QC & R

> **"Develop my memory and self-awareness!** ...do not negate myself ...or any aspect of myself! ...my mind is constantly aware of matters all around ...yet, I dismiss most of that ...and remember very little, ...does it seem wise to continue these habits? ...especially since I constantly dismiss my eternalself as well?..." Cotting, R. B., (2012) P. 14a

Write a couple sentences about me doing my eternalbeing actions today and what I desire that to look like!? Have an aware day whether it looks like my picture or not. Do not discontent myself. ;-)

EVENING PRACTICE:

BATHWAVE TIME FOR HEART ATTITUDES:	
BATHWAVEs **Part 1: Face & Embrace**	**Quantum Shifts** **Part 2: Replace with Grace**
I face and embrace my love letter that I avoid myself.	I can, I will and I am replacing my love letter with... I communicate with myself and celebrate without judgement.

Write down the present tense only of Part 2's in my list and say QC & R.

DAY 8:
GIVING MY VALUES HEAT

MORNING PRACTICE: I know how to start my day...DW, QC, & R.

Practice watching people!

> "...meeting my eternalself as a first-time experience will be of little value or significance if you aren't practiced in observing individuals! ...of course, in meeting my eternalself, you will be observing myself from a perspective you've not seen in a long time, so you need practice seeing myself! ...besides, carefully observing others is a worthy ability to develop! ...for it will strengthen my ability to access the archive of my body's (DNA)-data•information!" Cotting, R. B., (2012) P. 14a

Today be aware that when I begin to judge others I am not "observing". Think of myself as a camera for this exercise. I am just taking pictures of people and situations. Make it fun and actually ask to take pictures for a project I gave myself!

See if I can feel how much more energy I have in observing without judging?!

EVENING PRACTICE:

BATHWAVE TIME FOR HEART VALUES:	
BATHWAVEs **Part 1: Face & Embrace**	**Quantum Shifts** **Part 2: Replace with Grace**
I face and embrace my love letter that I deny the value of myself and others.	I can, I will and I am replacing my love letter acknowledging the value of my feelings and needs and others'.

Write down the present tense only of Part 2s in my notebook say QC, & R before bed.

DAY 9:
LOVE LETTER OF GRATITUDE NO MATTER WHAT I DO

MORNING PRACTICE: I start my day doing DW, QC, & R.
Reread yesterday and do the objective observing of people and situations today again without judging. I am taking real photo pictures whenever possible.

EVENING PRACTICE:

BATHWAVE TIME FOR HEART EMOTIONS:	
BATHWAVEs **Part 1: Face & Embrace**	**Quantum Shifts** **Part 2: Replace with Grace**
I face and embrace my love letter that I hate what I do.	I can, I will and I am replacing my love letter with I am filled with joy and gratitude no matter what I do.

Write down the present tense only of Part 2s in my list and say QC, & R.

DAY 10:
YOUR ULTIMATE PARTNER AND INTIMATE RELATIONSHIP

MORNING PRACTICE: I know what to do, my DW, QC, & R.

> **"Practice considering and describing the partner and intimate relationship you would choose to have for a lifetime**! ...be very specific about a partner's desirable talents and abilities, beliefs and attitudes, and actions and activities you desire ...and write everything down! ...and since a partner you would desire for a lifetime is much the same partner you would seek to communion with eternally, ...you might want to write of this to my eternalself! Cotting, R. B., (2012) #22 in Pericope 14a

Relate to everyone today as if this partner just spent the night with you loving you as you desire.. Don't tell anyone about it, yet smile big and imagine it as if it happened. LIKE THE FIRST TIME YOU FELL IN LOVE. Keep that awareness all day.

EVENING PRACTICE:

Write about my day and tonight feel the presence as it is true that my eternalself sleeps with me every night cherishing me as I desire. Treat myself like a special person safe and appreciated. I can kiss my own hand like I would a prince or princess, king or queen BECAUSE I AM!
Make sure I add all of my new replacements to my previous replacements and say all along with my QC.

QUANTUM REPLACEMENTS FOR MY FIRE ELEMENT HEART

I will repeat these present tense replacements for at least 21 days to 3 months (highly recommended). Before bed say with eyes closed. In the morning say with eyes open.

I live my eternal authentic and real self.
I know everything is a sacred mirror doorway of my whole self.

I am aware of my whole self.
I am love my aliveness unconditionally.
I am awakening my whole self.
I am safe and secure no matter what's happening.
I am appreciating how alive I am.
I listen to, acknowledge and communicate the feelings and needs of my friends.
I am having uplifting, complimentary thoughts of myself and then others.
I am open and aware of myself.
I am bringing myself anew to each relationship and myself in each moment.
I am reaching out to communicate in relationship.
I am appreciating the power of my entinglement.
I support resplendence with my communication.
I am connecting and conversing easily with people.
I speak my feelings and needs honestly to others.
I communicate with myself and celebrate without judgement.
I am acknowledging the value of my feelings and needs and others'.
I am filled with joy and gratitude no matter what I do.

Entanglement with things not known
Access through DNA

DECADE 12

SMALL INTESTINE FIRE MOTION

QUANTUM WAVES DANCE ENTINGLEMENT

My quantum waves
Entingle and motion my new
World as freeing dance.

DAY 1
QUANTUM CONVERSATION IS EXPANSIVE WITH DECLARATIVE QUESTIONING

MORNING PRACTICE: I know what to do as I complete my DW, QC, & R*.

Small Intestine motion:
This is about being nurtured by the quantum wave of living my illuminatedbeing. We will be exploring quantum communication as a source of self-nurturing since connecting in creative ways is a way of nurturing oneself.

Quantum Communication:

Resplendent Conversation
in an Emergent Entinglement Wave

"Understanding waves
To have the quality of our quantum "wave" nature in my conversations I need to practice qualities of the wave.

Quantum Waves keep on going to the ends of the Universe and then return to center, spin in order to go out again.
They are expansive in nature.
They never diminish or are interrupted.
They are transmitted to all beings.

The most important quality is that the **conversation keeps going and is expansive in nature.**

Quantum Questioning:
Questioning keeps a conversation open ended. Questions are truly expansive. **Questions need to be in the form of a declarative question so that I may declare myself at the same time as questioning. This type of questioning also does not demand an answer.**

Giving only an answer closes off the glow of communication, especially "yes and no". Answers can be narrowing and limited. Only answering restricts a conversation; even my own answers are not only restrictive, but my answers are all difficult to trust.

Truths are the worst offenders of all. It's insane that I look for truths. How do I know what I am saying is true? Let me give you an example of a conversation about this topic of searching for truth that uses declarative questioning.

Pat: "Doesn't that really mean then that any kind of search for truth is a foolish endeavor?"
Raj: "Is it possible that maybe a great deal of my life is based on foolish endeavors then?"
Pat: "Could there have been ways to live my life in a different way?"
Raj: "Wouldn't my life be entirely different had I used declarative questioning?"
Pat: "Might there be all kinds of ways that I might incorporate that into my life today?"
Raj: "Is it possible that maybe you have the choice to do anything you desire to do and be?"
Pat: "It seems to follow doesn't it that choices are wide open to me in this moment?"
Raj: "Consider that the universe will support you in any choice you make by not getting in the way?"
Pat: "Does that mean that any choice I make is not narrowing or restricting me or limiting me?"
Raj "Where in the universe does it ask that you make a specific choice that is good, better or best?"
Pat: "Does that mean that I can make any choice I desire for the first time in my life?"
Raj: "How different would my life be if you understood that life isn't judging you? You are judging you using life to do it at times. Can you stop doing that?"
Pat: "Wouldn't I be free to ask myself to be any way I wish?"

And so the conversation goes.

Can I see how the questions opened up a whole expansive realization on both sides of the conversation? This is the part of the conversation where I might start facing and embracing my self judgment and any emotion like anger or sadness as if I were Pat...

It is very important that I don't ask a yes or no question. A yes or no question closes the conversation as soon as it's asked. Using declarative questioning takes practice, but it is quite exciting to me that I can finally communicate in a way that feels very creative and empowering to both people in the conversation." Longo, A. (2012)

Just notice my conversation style today. Take a few notes about what I notice and practice asking a few questions that are not asking for yes or no answers.

EVENING PRACTICE: NONVIOLENT (NONJUDGMENTAL) COMMUNICATION

I often listen to Marshal Rosenberg on YouTube. He devised a method for nonviolent nonjudgmental communication (NVC) different than this, to use when I am in public. When possible, listen to one of his talks.

If not reread the morning reading; there so much to learn from it!
Before sleep ask for a dream about my communication style. Am I asking questions that enhance the wonderment of living self NURTURING illumination?!
Do quantumcode, replacements keeping my eyes closed.

DAY 2
SMALL INTESTINE BELIEF PATTERNS

MORNING PRACTICE: Do quantumcode, replacements with my eyes open.

Write down some beginnings of questions on a card or paper I can ask today?
Here are some examples:

- Have I considered…?
- How different might it be if...?
- Wouldn't I be free to ask myself...?
- Might it be that....?
- Is it possible that....?
- It seems to follow doesn't it...?

EVENING PRACTICE:

BATHWAVE TIME FOR SMALL INTESTINE BELIEFS:	
BATHWAVEs **Part 1: Face & Embrace**	**Quantum Shifts** **Part 2: Replace with Grace**
I face and embrace my love letter of believing that nothing is nourishing in my life.	I can, I will and I am replacing the old belief with perceiving that all my mirror-calls nourish me.

Write down the present tense only of Part 2s in my notebook and say my quantumcode and replacements with my eyes closed.

DAY 3
SMALL INTESTINE ACTION PATTERN REPLACEMENTS

MORNING PRACTICE: Say my quantumcode and replacements keeping my eyes open.
I practice using my declarative questions in conversations today.

EVENING PRACTICE:

BATHWAVE TIME FOR SMALL INTESTINE ACTIONS:	
BATHWAVEs **Part 1: Face & Embrace**	**Quantum Shifts** **Part 2: Replace with Grace**
I face and embrace my love letter: that I block my nurturance as I live.	I can, I will and I am living that everything nurtures me.
I face and embrace my love letter: that I interpret the actions that I encounter as hindrances or barriers.	I can, I will and I am appreciating all I encounter as a mirror-call.

Write down the present tense only of Part 2s in my notebook. Recite my quantumcode and replacements with my eyes closed.

DAY 4
SMALL INTESTINE THOUGHT REPLACEMENTS

MORNING PRACTICE: Say my quantumcode and replacements remembering to keep my eyes open.
Continue with using the card with questions after my responses.

EVENING PRACTICE:

BATHWAVE TIME FOR SMALL INTESTINE THOUGHTS:	
BATHWAVEs **Part 1: Face & Embrace**	**Quantum Shifts** **Part 2: Replace with Grace**
I face and embrace my love letter: that I am distracting myself from the fullness of the moment.	I can, I will and I am replacing my love letter with my mirror-call that I am attracting everything I desire in the fullness of the moment.

Write down the present tense only of Part 2s and speak my quantumcode and replacements with my eyes closed.

DAY 5
FULLY PRESENT FOR SMALL INTESTINE HABITS

MORNING PRACTICE: FOCUS MYSELF IN EACH MOMENT
Do my quantumcode and replacements keeping my eyes open.
If this is difficult then do a bathwave to make it possible. The BATHWAVE will be about judging myself or them and emotions like boredom that come from it?

EVENING PRACTICE:

BATHWAVE TIME FOR SMALL INTESTINE	
BATHWAVEs **Part 1: Face & Embrace**	**Quantum Shifts** **Part 2: Replace with Grace**
I face and embrace my love letter: that I always negate my feelings.	I can, I will and I am replacing my love letter with I nourish myself with feelings I desire.
I face and embrace my love letter that I look for affirmation from others	I can, I will and I am replacing my love letter with I affirm and celebrate my communion of self.

Write down the present tense only of Part 2s and recite my quantumcode and replacements keeping my eyes closed.

DAY 6:
USING NURTURING CREATIVE WORDS

MORNING PRACTICE: After dreamwaving(DW) say my quantumcode(QC) and replacements(R) remembering to keep my eyes open.

Include my quantumcode in conversations by asking my self what would my abidingself like to say and then using my card find a way to say it with a question? Friends or family are probably good choices to practice declarative questioning with.

EVENING PRACTICE:

BATHWAVE TIME FOR SMALL INTESTINE WORDS:	
BATHWAVEs **Part 1: Face & Embrace**	**Quantum Shifts** **Part 2: Replace with Grace**
I face and embrace my love letter that my words justify, excuse and judge me or others.	I can, I will and I am replacing my love letter with my words are emergent spontaneously and creatively.

Write down the present tense only of Part 2s and recite my quantumcode and replacements keeping my eyes closed.

DAY 7
REPLACING SMALL INTESTINE'S PASSION

MORNING PRACTICE: Say my quantumcode and replacements remembering to keep my eyes open.
Do something different so all conversations nurture me like adding a smile.

EVENING PRACTICE:

BATHWAVE TIME FOR SMALL INTESTINE ATTITUDES:	
BATHWAVEs **Part 1: Face & Embrace**	**Quantum Shifts** **Part 2: Replace with Grace**
I face and embrace my love letter that I have lost my desire to cook and live.	I can, I will and I am replacing my love letter with I cook and live with passion.

Write down the present tense only of Part 2's and say QC & R before bed.

DAY 8
EVERYTHING NOURISHES ME AS I ADD MY MEANING

MORNING PRACTICE: Say my quantumcode and replacements remembering to keep my eyes open.
Be aware of passion returning today no matter what I do..

EVENING PRACTICE:

BATHWAVE TIME FOR SMALL INTESTINE VALUES:	
BATHWAVEs **Part 1: Face & Embrace**	**Quantum Shifts** **Part 2: Replace with Grace**
I face and embrace my love letter that life is meaningless or draining.	I can, I will and I am replacing my love letter with everything nourishes me and gives me the meaning I choose.

Write down the present tense only of Part 2s in my notebook then state QC & R. Ask for any dream informotion you desire.

DAY 9
GIVING MY QUANTUMCODE MEANING TO EVERYTHING I DO TODAY

MORNING PRACTICE: Record my dreams for DW. Say my quantumcode and replacements remembering to keep my eyes open.

Remember my quantumcode no matter what I do or think and see if I can bring the meaning that is important to me in the situation or interaction.

EVENING PRACTICE:

BATHWAVE TIME FOR SMALL INTESTINE EMOTIONS:	
BATHWAVEs **Part 1: Face & Embrace**	**Quantum Shifts** **Part 2: Replace with Grace**
I face and embrace my love letter that I am starving for warmth and nourishment in my life.	I can, I will and I am self- sustaining and the source of my own rejuvenation.

Write down the present tense only of Part 2s in my notebook then state QC & R. Ask for a dream about how to do this?

DAY 10
IS EVERYTHING NURTURING ME IN THIS MOMENT?

MORNING PRACTICE: CAN I SHARE SELF-NURTURING?
Do dreamwaving first.
Say my quantumcode and replacements remembering to keep my eyes open.

Do automatic writing about sharing self-nurturing using the 'Airplane Principle'. This means putting the oxygen mask on myself first even if a child is sitting next to me. Do I know why? After writing see how much I can transform my day this way! Picture it being as I desire.
Take notes, as I will need to write about my day.

EVENING PRACTICE:
Write about how I am noticing that my day was NURTURING me with my emergent miracles. I can embrace those that I would like to replace.

Do QC & R with my eyes closed.

QUANTUM REPLACEMENTS FOR MY FIRE ELEMENT
SMALL INTESTINE

I will repeat these present tense replacements for at least 21 days to 3 months (highly recommended). Before bed say with eyes closed. In the morning say with eyes open.

I am perceiving that all my mirror-calls nourish me.
I am living that everything nurtures me.
I am appreciating all I encounter as a mirror-call.
I am attracting everything I desire in the fullness of the moment.
I nourish myself with feelings I desire.
I affirm and celebrate my communion of self.
I my words are emergent spontaneously and creatively.
I cook and live with passion.

Everything nourishes me and gives me the meaning I choose.
I am self- sustaining and the source of my own rejuvenation.

DW, QC and R.= After dreamwaving(DW) say my quantumcode(QC) and replacements(R) remembering to picture what it looks like living each as I say one.

DECADE 13

PERICARDIUM FIRE MOTION DECADE

DAY 1
PERICARDIUM MOTION:

MORNING PRACTICE: After dreamwaving(DW) do quantum code & replacements with my eyes open.

Bonding is the motion of the pericardium, which is the water balloon that the heart beats inside. Bonding is a basic need. When we deny it, we sometimes suffer illnesses. I once read a meaningful analogy in a book, which was about how a hand held trampoline represents life. Its point was that for a human being to thrive we need a minimum of three significant relationships to hold the edges of our trampoline or we are not resilient. More people holding the trampoline is desirable.

This is the importance of bonding, Ashley Montague writes, "Babies die when not touched and kept in a nursery" (Montagu, 1986). This may be a contributing constituent of dying in the aged also or anyone for that matter.

This gives hugging a great significance. How many hugs can I give today?
Observe any unmet needs I have in regards to connecting and bonding.

EVENING PRACTICE:
Write the unmet needs I observed today in myself. Check the list of needs in the appendix if I can't name them.

Jot down some of the BATHWAVEs I would like to add so I can meet these unmet needs to warm myself up?

Entinglement:
Picture myself living this way regardless of others attitudes or responses. Imagine extending warmth, connection, and bonding to everyone tomorrow with a smile and offer even more when possible, like a hug. Please do it without judging that it is good or bad. I can ask to give a hug. "May I give you a hug? I really need one." Be ok if I receive a no.

Say out loud my quantumcode & replacements with my eyes closed.

DAY 2
BEGINNING WITH PERICARDIUM BELIEFS

MORNING PRACTICE: Say my quantumcode & replacements with my eyes open.

Focus on the emergent nature of life. Only the information I amass in my life, comes together in emergent ways, which means in surprising ways. Sometimes, I question my mirror-calls, since I do not perceive the nature of the desirability. Sometimes I am not aware of all the information I live with. For instance, very few are aware of the content of their unconscious which is included in this process!

This is why this process of preparing me to cross the bridge and live the quantum way takes a few months to do. I comb the relevant information that forms the basis of most of my undesirable BATHWAVEs.

I then replace them by pouring into my life the new true ways I desire.

Be observant today of these emergent miracles, instead of the causing and effecting that my language is incorrectly structured upon.

EVENING PRACTICE:

BATHWAVE TIME FOR PERICARDIUM BELIEFS:	
BATHWAVEs **Part 1: Face & Embrace**	**Quantum Shifts** **Part 2: Replace with Grace**
I face and embrace my mirror-call that others affect me. I face and embrace I blame and accuse others.	I can, I will and I am replacing my misunderstanding of the nature of reality with my love letter of transforming what I see to appreciate and bless everyone.
I face and embrace my love letter of believing I am not free to be me.	I can, I will and I am replacing this with I am excited to be me as I freely desire.

I write down the present tense only of Part 2s then say my quantumcode and replacements before sleep.

DAY 3
BATHWAVE TIME FOR PERICARDIUM ACTIONS

MORNING PRACTICE: After DW, recite my quantumcode & replacements with my eyes open.

"Practice being everything you do! ...like it or not, language only has a place for you as the causer of things ...or the one to whom things happen! ...while life also has a place and life for you as the action itself ...and the issue of myself as an action! ...however, you have forsaken this expansive and extensive nature of life for the false comfort and security of language! ...does it seem wise to continue in this primitive choice and way of life? …are you interested in being and experiencing all you are?" Cotting, R. B., (2012), #20 in Pericope 14a

Keep a record of the things I do today to recall this evening when I transform my pericardium actions!

EVENING PRACTICE:

BATHWAVE TIME FOR PERICARDIUM ACTIONS:	
BATHWAVEs **Part 1: Face & Embrace**	**Quantum Shifts** **Part 2: Replace with Grace**
I face and embrace my love letter: that I reject my mother, father, nature and the world.	I can, I will and I am replacing my love letter that the universe supports me abundantly.
I face and embrace my love letter that I reject my mind, body, feelings and eternalself.	I can, I will and I am replacing my love letter that I am an eternal mindful being.
I face and embrace my love letter of that I am needy for my mother, father, nature, and or the world's approval.	I can, I will and I am replacing my love letter with I nurture and guide myself without judgement.
I face and embrace my love letter of being needy for my mind, body, feelings or being codependent with others.	I can, I will and I am replacing my love letter with my mirror-call of: I am living as a self sustaining resplendent communion of mindful being.

Write down the present tense only of Part 2s and say my QC & R before bed.

DAY 4
BATHWAVE TIME FOR PERICARDIUM THOUGHTS:

MORNING PRACTICE: Say my quantumcode & replacements with my eyes open.
Reread and follow directions from yesterday's morning practice and do again!

EVENING PRACTICE:

BATHWAVE TIME FOR PERICARDIUM THOUGHTS:	
BATHWAVEs **Part 1: Face & Embrace**	**Quantum Shifts** **Part 2: Replace with Grace**
I face and embrace my love letter: that think I am not sexy enough.	I can, I will and I am replacing my love letter I am a vital, sexy being.
I face and embrace my love letter that I accuse others for my difficulties or problems.	I can, I will and I am replacing my love letter I appreciate everything is a sacred doorway for my illuminated being.

I face and embrace my love letter that I blame others for my difficulties or problems.	I can, I will and I am replacing my love letter I bless others as sacred doorways for my illuminated being.
I face and embrace my love letter that I complain about others or myself and discontent myself.	I can, I will and I am replacing my love letter I connect and communicate as my illuminated being without discontenting myself.
I face and embrace my love letter that difficulties and problems exist.	I can, I will and I am replacing my love letter with everything is my mirror-call for my universal illumination.
I face and embrace my love letter that I think that things are opposing me in the universe.	I can, I will and I am replacing my love letter with everything supports me as I support myself.
I face and embrace my love letter that the thought of failure motivates me.	I can, I will and I am replacing my love letter that no matter what occurs I am successful, sharing and inspiring my new self.

Write down the present tense only of Part 2s and say my QC & R before bed.

DAY 5
BATHWAVE TIME FOR PERICARDIUM HABITS

MORNING PRACTICE: Start my morning with my QC & R.

> "Begin observing and feeling myself! ...feel my personalself communioned with my body! ...listen carefully to what you are truly saying ...not merely to the words! ...and practice stepping away from whatever you are doing to perceive myself! ...don't just watch what you are doing!
>
> You've probably felt sick ...and even felt pleased and happy, ...but have you ever felt myself feeling sick or pleased or happy? ...and if you think carefully about it, can't you observe myself being sick or pleased ...or happy?
>
> Practice feeling myself! ...not just feeling my body ...or feeling the same as my body feels! ...feel the self inside my body! ...then practice watching that self within my body!
>
> When you hurt myself, did you do that to myself ...or to my body? ...and who saw you do that? ...are you willing to give my full attention to everything you are doing and saying at every moment? ...and simultaneously pay close attention to you ...saying and doing? ...you are my own witness! ...what would you like to see myself doing ...and saying?" Cotting, R. B., (2012) Pericope 14a.

EVENING PRACTICE:

BATHWAVE TIME FOR PERICARDIUM HABITS:	
BATHWAVEs **Part 1: Face & Embrace**	**Quantum Shifts** **Part 2: Replace with Grace**
I face and embrace my love letter: that I am guilty.	I can, I will and I am replacing my love letter that I am innocent.
I face and embrace my love letter that I am resentful as I live.	I can, I will and I am replacing my love letter with I desire to communicate a wonderful life to everyone I meet.
I face and embrace my love letter that I am overbearing in what I do.	I can, I will and I am replacing my love letter I listen and empathise with the feelings and needs of others.
I face and embrace my love letter that I use force, control and power to manipulate to get things done.	I can, I will and I am replacing my love letter with as I transform myself and watch others and things transform around me.

Write down the present tense only of Part 2s and say my QC & R before bed.

DAY 6
BATHWAVE TIME FOR PERICARDIUM WORDS

MORNING PRACTICE: Do quantumcode & replacements with my eyes open.
Reread and follow directions from yesterday's morning practice and do the practice again!

EVENING PRACTICE:

BATHWAVE TIME FOR PERICARDIUM WORDS:	
BATHWAVEs **Part 1: Face & Embrace**	**Quantum Shifts** **Part 2: Replace with Grace**
I face and embrace my love letter that my words are alienating and isolating.	I can, I will and I am replacing my love letter that.. my words express the union of communion of the quantum way.
I face and embrace my love letter that others words trigger me.	I can, I will and I am replacing my love letter that I unconditionally listen to others ideas, feelings and needs.

Write down the present tense only of Part 2s and say my QC & R before bed.

DAY 7
BATHWAVE TIME FOR PERICARDIUM ATTITUDES

MORNING PRACTICE: Say my quantumcode & replacements with my eyes open.

> **"Develop my memory and self-awareness!** ...do not negate myself ...or any aspect of myself! ...my mind is constantly aware of matters all around ...yet, you dismiss most of that ...and remember very little, ...does it seem wise to continue these habits? ...especially since you constantly dismiss my eternalself as well?..." Cotting, R. B., (2012) P14a

Write a couple sentences about me doing my eternalbeing actions today and what I desire that to look like! Have an aware day whether it looks like my picture or not do not discontent myself. ;-)

EVENING PRACTICE:

BATHWAVE TIME FOR PERICARDIUM ATTITUDES:	
BATHWAVEs **Part 1: Face & Embrace**	**Quantum Shifts** **Part 2: Replace with Grace**
I face and embrace my love letter that my negative attitudes block my clarity and strength of self.	I can, I will and I am replacing my love letter with my true strength is my personal change.
I face and embrace my love letter that I doubt my true strength and wisdom.	I can, I will and I am replacing my love letter with I have full faith in my personal change.

Write down the present tense only of Part 2s and say my QC and R before I sleep.

DAY 8
BATHWAVE TIME FOR PERICARDIUM VALUES

MORNING PRACTICE: Start my morning with my quantumcode & replacements.

"Kindness in words
        creates confidence.
Kindness in thoughts
        creates profoundness.
Kindness in giving
        creates love."
Lao Tzu ancient Chinese philosopher.

I practice being kind to myself today. I am replacing any words, thoughts and actions that are not uplifting to myself and I am generous with myself.

As Shakespeare wrote: "Be true to yourself and it shall follow as the night, the day, you canst be false to anyone." Kindness is being true or it isn't kind?

EVENING PRACTICE:

BATHWAVE TIME FOR PERICARDIUM VALUES:	
BATHWAVEs **Part 1: Face & Embrace**	**Quantum Shifts** **Part 2: Replace with Grace**
I face and embrace my love letter that there is no free will.	I can, I will and I am replacing my love letter with I value my freewill to keep or recycle whatever I desire.
I face and embrace my love letter that the universe is out to get me.	I can, I will and I am replacing my love letter with the universe supports me as I support myself.

Write down the present tense only of Part 2s and say my QC and R before I sleep.

DAY 9
BATHWAVE TIME FOR PERICARDIUM EMOTIONS

MORNING PRACTICE: Say my QC & R with eyes open.

Reread yesterday and do the objective observing of people and situations today again without judging. I am taking real photo pictures where possible.

EVENING PRACTICE:

BATHWAVE TIME FOR PERICARDIUM EMOTIONS:	
BATHWAVEs **Part 1: Face & Embrace**	**Quantum Shifts** **Part 2: Replace with Grace**
I face and embrace my love letter that fear of failure paralyzes me.	I can, I will and I am replacing my love letter with I understand that there are no mistakes or failures. I can, I will and I do never give up.
I face and embrace my love letter that I feel disoriented and out of my orbit.	I can, I will and I am replacing my love letter with I am fired up with enthusiasm.

Write down the present tense only of Part 2s and say my QC and R before I sleep.

DAY 10
EVERYONE IS A RESPLENDENT CONSTITUENT OF MY LIVING

MORNING PRACTICE: After DW, outloud say good morning to my quantumcode & replacements with my eyes open.

> **Practice considering and describing the partner and intimate relationship you would choose to have for a lifetime**! ...be very specific about a partner's desirable talents and abilities, beliefs and attitudes, and actions and activities you desire ...and write everything down! ...and since a partner you would desire for a lifetime is much the same partner you would seek to communion with eternally, ...you might want to write of this to my eternalself!" Cotting, R. B. (2012) Pericope 14a #22.

Relate to everyone today as if this person just spent the night with me loving me as I desire. Not only as in making love. It could be a dinner or at a party or in a class or on a hike. Don't tell anyone about it, yet smile big and imagine it as if it happened. Keep that feeling with me all day.

EVENING PRACTICE:
MAKE A NEW RECORDING of the replacements from Decades I have not recorded yet.
Or just organize all my pages to read. Congratulations, I am almost complete with my quantum bridge work!
Say my QC and R out loud before I sleep.

QUANTUM REPLACEMENTS FOR MY FIRE ELEMENT
PERICARDIUM

I will repeat these present tense replacements for at least 21 days to 3 months (highly recommended). Before bed say with eyes closed. In the morning say with eyes open.

I am transforming what I see to appreciate and bless everyone.
I am excited to be me as I freely desire.
The universe supports me abundantly.
I am an eternal mindful being.
I nurture and guide myself without judging.
I am living as a self-sustaining resplendent communion of mindful being.
I am a vital, sexy being. I appreciate everything is a sacred doorway for my illuminated being.
I bless others as sacred doorways for my illuminated being.
I connect and communicate as my illuminated being without discontenting myself.
Everything is my mirror-call for my universal illumination.
Everything supports me as I support myself.
No matter what occurs I am successful, sharing and inspiring my new self.
I am innocent.
I desire to communicate a wonderful life to everyone I meet.
I listen and empathize with the feelings and needs of others.

I transform myself and watch others and things transform around me.
My words express the union of communion of the quantum way.
I unconditionally listen to others ideas, feelings and needs.
My true strength is my personal change.
I have full faith in my personal change.
I value my freewill to keep or recycle whatever I desire.
The universe supports me as I support myself.
I understand that there are no mistakes or failures.
I never give up.
I am fired up with enthusiasm.

After DW, say QC and R.= After dreamwaving(DW), say my quantumcode(QC) and replacements(R) remembering to picture what it looks like living each one as I say it.

DECADE 14

TRIPLE HEATER FIRE MOTION

DAY 1
TRIPLE HEATER MOTION IS THAT EVERYTHING IS IN HARMONY WITH ME

MORNING PRACTICE: After dreamwaving(DW), say my quantumcode(QC) and replacements(R) remembering to picture what it looks like living each one as I say it.

IT IS NO ACCIDENT THAT THERE ARE THREE HEATING SPACES IN MY BODY AS A TRIUNITY AND THREE TRIANGULAR BONES TO DESIGNATE EACH OF THESE SPACES: SACRUM, STERNUM, AND JAW.

> "The greatest meaning of life is not that of serving others ...it is offering to life! ...for in that, you offer to myself! ...do everything as an offering to life ...and my life and understandings will change substantially ...and immediately! ...but only if there are no special or compassionate offerings, no right or divine offerings, and no actions of service ...all actions are my offerings! ...and do not treat communioning with my eternalself as anything but my offering to the world around you!
>
> An expansive and extensive expression of myself as an eternalbeing requires a new and different agreement with the world around you! ...in the past, my agreement was that the universe would be to you as you are to the universe! ...and from that, you felt you would be rewarded for right thinking and good deeds! ...that you would be loved as you love others ...that it will be done unto you as you do unto others! ...and you would be served as you serve others! ...all of this was unfortunately expressed as, "Do unto others as you would have others do unto you"!
>
> Certainly, you are now willing to change my agreement with life ...such that the more you offer ...the more you will discover you have to offer! ...the more you communion ...the more you will be communioned! ...that service to others is actually service to myself! ...and you are expanded only by what you communion with ...such that you have my eternalself and life only as you give life to my eternalself ...and communion that with my personalself and life!
>
> Every act of life is complete in itself! ...there is no reward or punishment! ...no return for actions! ...and no quid pro quo! ...for what you communion is what you are ...and what you live and have! ...you do not reap what you sow ...you live what you sow! ...in fact, though you don't seem to realize it, you are what you sow! ...and everything you sow is an offering to life. Cotting, R. B. (2012) P14a

EVENING PRACTICE:
Contemplate in a quiet state how my life is in harmony with ME always. Get a picture of the nature of my life today. Is this how I want my living to continue? When the answer is no, embrace it and replace it writing down the new desire.

The triple heater gives me the ability to see.

Write down the present tense quantum shifts and speak my quantumcode & replacements keeping my eyes closed.

DAY 2
TRIPLE HEATER BELIEF REPLACING

MORNING PRACTICE: After DW, do my quantumcode & replacements with my eyes open. Reread Day 1 and notice what you sow during the day.

EVENING PRACTICE:

BATHWAVE TIME FOR TRIPLE HEATER BELIEFS:	
BATHWAVEs **Part 1: Face & Embrace**	**Quantum Shifts** **Part 2: Replace with Grace**
I face and embrace my love letter of believing that my world is not in harmony with me.	I can, I will and I am replacing the past with my mirror-calls of living are in harmony with me, for my transformation.
I face and embrace my love letter: that I believe that I should judge.	I can, I will and I am replacing my love letter that good and bad do not exist I am willing to reflect on the mirror-call of what is really going on in me.

Write down the present tense quantum shifts and speak my quantumcode & replacements keeping my eyes closed.

DAY 3
TRIPLE HEATER ACTIONS

MORNING PRACTICE: Do my quantumcode & replacements with my eyes open.

Reread Day 2 and notice what I sow during the day.

EVENING PRACTICE:

BATHWAVE TIME FOR TRIPLE HEATER ACTIONS:	
BATHWAVEs **Part 1: Face & Embrace**	**Quantum Shifts** **Part 2: Replace with Grace**
I face and embrace my love letter: that I am disoriented and lost.	I can, I will and I am living my eternalself in each moment.

Write down the present tense quantum shifts and speak my quantumcode & replacements keeping my eyes closed.

DAY 4
TRIPLE HEATER THOUGHTS

MORNING PRACTICE: Speak/sing my quantumcode & replacements with my eyes open.

> **"Practice being everything you do!** ...like it or not, language only has a place for you as the causer of things ...or the one to whom things happen! ...while life also has a place and life for you as the action itself ...and the issue of myself as an action! ...however, you have forsaken the expansive and extensive nature of life for the false comfort and security of language! ...does it seem wise to continue in this primitive choice and way of life? ...are you interested in being and experiencing all you are? Cotting, R. B. (2012)

EVENING PRACTICE:

BATHWAVE TIME FOR TRIPLE HEATER THOUGHTS:	
BATHWAVEs **Part 1: Face & Embrace**	**Quantum Shifts** **Part 2: Replace with Grace**
I face and embrace my love letter: that my life is not in harmony with me.	I can, I will and I am replacing my love letter with my mirror-call that everything is in "entinglement" with me therefore it's always in harmony with me.

Write down the present tense quantum shifts and speak my quantumcode & replacements keeping my eyes closed.

DAY 5
BATHWAVE TIME FOR TRIPLE HEATER HABITS

MORNING PRACTICE: After DW, do my quantumcode & replacements with my eyes open. Reread Day 4 and take note of what I notice during my day.

EVENING PRACTICE:

BATHWAVE TIME FOR TRIPLE HEATER HABITS:	
BATHWAVEs **Part 1: Face & Embrace**	**Quantum Shifts** **Part 2: Replace with Grace**
I face and embrace my love letter: that I am always trying to change the world.	I can, I will and I am replacing my love letter with my mirror-call that as I change myself I am changing the world.

DAY 6
BATHWAVE TIME FOR TRIPLE HEATER WORDS

MORNING PRACTICE: Do my quantumcode & replacements with my eyes open.

> "Decide and declare, if only to myself, that you are meeting and experiencing my eternalself in every event and action! …begin every situation and every moment of my life with a firm declaration of who you are …and what you are doing! …not what you want to do or will do …or what you intend to do! …and certainly not who you hope to become! …and do not begin with an "open mind" …that you accept what comes to you! …for everything in my life depends upon what you declare and commit to …not on what you accept or allow!
>
> Do not struggle …or stress myself to fulfill my decisions and declarations, …remain assured …and do not discontent myself! …and carry my promises and commitments with you in my daily affairs …in everything you do!" Cotting, R. B., (2012).

EVENING PRACTICE:

BATHWAVE TIME FOR TRIPLE HEATER WORDS:	
BATHWAVEs **Part 1: Face & Embrace**	**Quantum Shifts** **Part 2: Replace with Grace**
I face and embrace my love letter that my words are cold and judgmental.	I can, I will and I am replacing my love letter with my words are warm and empathic.

In my notebook I write down the present tense quantum shifts and speak my quantumcode & replacements keeping my eyes closed.

DAY 7
BATHWAVE TIME FOR TRIPLE HEATER ATTITUDES:

MORNING PRACTICE: Say my quantumcode & replacements with my eyes open.
Reread Day 6 and what do I declare or commit to on this day?

EVENING PRACTICE:

BATHWAVE TIME FOR TRIPLE HEATER ATTITUDES:	
BATHWAVEs **Part 1: Face & Embrace**	**Quantum Shifts** **Part 2: Replace with Grace**
I face and embrace my love letter that I am depressed and despairing at this disharmony.	I can, I will and I am replacing my love letter with my mirror-call that I am full of elation as I dance with the glow of living.

Write down the present tense quantum shifts and speak my quantumcode & replacements keeping my eyes closed.

DAY 8
BATHWAVE TIME FOR TRIPLE HEATER VALUES

MORNING PRACTICE: Do my quantumcode & replacements with my eyes open.

"Practice making connections! ...to expand my view and image of my self and life ...and to understand that what you do in one area of my life you do in every area of my life ...practice making connections!" Cotting, R. B. (2012)

EVENING PRACTICE:

BATHWAVE TIME FOR TRIPLE HEATER VALUES:	
BATHWAVEs **Part 1: Face & Embrace**	**Quantum Shifts** **Part 2: Replace with Grace**
I face and embrace my love letter that my feminine side does not value my masculine side.	I can, I will and I am replacing my love letter with that my feminine side (feeling, poetic, being side) values my masculine side (linear, logical thinking, organized side).

I am writing down the present tense quantum shifts and speak out loud my quantumcode & replacements with my eyes closed.

DAY 9
BATHWAVE TIME FOR TRIPLE HEATER EMOTIONS

MORNING PRACTICE: Do my quantumcode & replacements with my eyes open.
Reread Day 8 and practice making connections; notice how my areas of life connect.

EVENING PRACTICE:

BATHWAVE TIME FOR TRIPLE HEATER EMOTIONS:	
BATHWAVEs **Part 1: Face & Embrace**	**Quantum Shifts** **Part 2: Replace with Grace**
I face and embrace that I hate the disharmony of relationships.	I can, I will and I do delight in stimulating relationships.
I face and embrace my love letter that I close my energy down in relationships.	I can, I will and I am energizing myself in relationship.

Write down the present tense quantum shifts and speak my quantumcode & replacements keeping my eyes closed.

DAY 10
PRACTICE SEEING OTHERS AS ETERNALBEINGS HELPS ME EXPERIENCE MY RESPLENDENTBEING

MORNING PRACTICE: Recite my quantumcode & replacements with my eyes open.

> **Practice seeing others as eternalbeings!** ...seeing myself with ordinary eyes keeps you from seeing myself as extraordinary ...or eternal! ...and seeing myself with extraordinary eyes that are normal for you ...certainly keeps you from seeing myself as extraordinary! ...in the same way, you cannot see or know myself as eternal while eternal! ...therefore, seeing others as eternalbeings enables you to know and experience myself as eternal!
>
> Know myself and you will not actually know much about myself! ...for knowing myself is too self-referential to be of any real value! ...however, truly and genuinely knowing my neighbor as an eternalbeing ...and you as a resplendentbeing will become obvious! Cotting, R. B., (2012)

Today be aware of how I am viewing other people. Can I do it without judging words?

EVENING PRACTICE:
Write a few lines about how I did today with my practice without using any judging words or ideas of good and bad or right or wrong.
Ask for a dream to help me understand how I am viewing people.
Make a new recording of all replacements since the last. Then take a moment and just marvel at all the transformation I have invited into my living.

Before I sleep speak my quantumcode and replacements closing my eyes.

Congratulations I am fired up and ready to melt my age-bergs and more!

THE YANG SIDE OF FIRE: Chapter 38 From the Tao Te Ching: reworded by A. Longo
A truly good person is not aware of their goodness,
And is always beginning anew.
A foolish man "tries" to be good
And is therefore not good.
When one "tries" to do anything
Our life becomes "trying".

A truly real person emerges without dissecting the "thinking" of doing.
That leaves nothing incomplete (undone).
A foolish person is always "trying" to do
Yet much remains to be done.

When a disciplinarian does something and no one responds,
They roll up their sleeves in an attempt to enforce order.

Therefore, when the Tao (unconditional living, which is emergent) is lost, goodness is born.
When goodness is lost, trying to be kind is born.
When "trying to be kind" is lost, justice is born.
When justice is lost, ritual is born.
Now ritual is a mask for faith and loyalty and the beginning of confusion.
Knowledge of the future is the beginning of foolishness and only a flowery trapping of my emergent nature of my true self.

Therefore, the truly great man dwells on what is real (authentic)
And not what is on the surface,
On the fruit and not the flower.
Therefore, support living an emergent surprise.

Transform my "trying"

QUANTUM REPLACEMENTS FOR MY FIRE ELEMENT
TRIPLE HEATER

I will repeat these present tense replacements for at least 21 days to 3 months (highly recommended). Before bed say with eyes closed. In the morning say with eyes open.

My mirror-calls of living are in harmony with me, for my transformation.
Good and bad do not exist...I am willing to reflect on the
mirror-call of what is really going on in me.
I am living my eternalself in each moment.
Everything is in "entinglement" with me therefore it's always in harmony with me.
I change myself I am changing the world.
My words are warm and empathic.

I am full of elation as I dance with the glow of living.
My feminine side (feeling, poetic, being side) values my masculine side (linear, logical thinking, organized side).
I do delight in stimulating relationships.
I am energizing myself in relationship.

After DW, say QC and R.= After dreamwaving(DW), say my quantumcode(QC) and replacements(R) remembering to picture what it looks like living each one as I say it.

DECADE 15

MELT AND REPLACE FROZEN AGE LIMITS, QUANTUM STYLE; and QUANTUM TOOLS OF ABUNDANCE

DAY 1
MELTING FROZEN AGE PATTERNS

Instantly Changing Conversations with Myself

Would I consider limiting myself to act as a 3 or 5 year old child? Or a 9 to 10 year old beloved? Or an 11 or 12 year old mother or father?
This is what I discovered about myself when I muscle tested the age of three aspects of myself: my child, beloved, and parent. Each of these three want progressing, yet they do not because it is an old habit that is comfortable, which stands in the way of the illuminatedself.

This is the reason progressing the past of the six aspects of my personalself as my father, mother, male and female children, and beloveds (representing my self-guidance, self-nuturance, male and female creativity, male and female self-love) is a priority after empowering the quantum abidingself.

Thanks to the understanding of quantum 'entanglement' between all the cells in our body I can identify the issues that froze me at various ages of my living. Next, I can replace them with what I desire relieving this old habit, giving me the ability to rewrite my past of outdated reruns and rewrite them any way I desire.

MORNING PRACTICE: After DW, remember to recite my quantumcode and replacements with my eyes open to start my morning.

Guess the overall frozen age of various people I interact with today. Notice my response to these people which may be an indicator of my own frozen overall age.

EVENING PRACTICE: CHOOSING WHICH ASPECT OF ME NEEDS REPLACING FIRST
This process is done by using noticing, muscle testing, the Googleverse Access Method Express (GAME), a pendulum, or any other method for getting an answer, including flipping a coin. Choose the method I desire. If using muscle testing see appendix.

Step 1: Using GAME or muscle testing look at figure 4.1 to determine which of my six aspects needs melting first.
Step 2: Now I have identified my aspect, I need to determine what age I am frozen at. If muscle testing use the same method as in step one beginning at birth counting up until an age says no. It is the year before that is determined.

Or if using my Googleverse Access Method Express (GAME) use the multiple of tens bag and single numbers bag together then I can discover an exact age. For example if I pull out a 30, I know my answer is in my 30s. I can then ask if I was 30 years of age. If I get a no, I can ask what age in my 30s? Then pull a number from the Numbers bag to receive my answer. If I pull out a 4, I am 34 years old.
Step 3: Ask myself what the issue was? I can use the questioning method from the six areas of my life or I could test the 5 Element BATHWAVES (both resources found in tomorrow's evening practice on Day 2). Replace the BATHWAVE I reveal during this process. If I am doing well with accessing all of this information then continue until I reach the age of 200 years old.
When accessing ages greater than my present age, ask about issues related to my mother's, father's, or grandparent's ages. Their issues will be my issues once identified I can replace them.
Step 4: I will continue this process for the next 5 days choosing another aspect of my life and repeat steps 2 and 3. This whole process can take an hour or more.
Step 5: When all six aspects are complete I should feel like a free individual whose weight has been lifted.
If I would like to work with a Quantum Life Coach, it is possible to develop a coaching relationship online to complete some of these techniques if needed. Please see website for a list of Quantum Life Coaches.

One Turkish client was so relieved afterward that he said, "You should do this for everyone in the world."

If I have not completed melting the frozen ages of my first aspect, do not worry. I can complete it in the morning.

Each evening write down my quantum replacements in the present tense. Then remember to recite my quantumcode and replacements with my eyes closed before sleep.

DAY 2
FINDING THE ISSUES OF ANY FROZEN AGE USING 6 AREAS OF LIVING AND/OR 5 ELEMENTS AS WE DID IN GAME

MORNING PRACTICE: Say my quantumcode and replacements with my eyes open.

During my day notice if I feel different with any of my relationships after completing last night's replacing my frozen age patterns. Write about it when I have time.

EVENING PRACTICE: Choose a new aspect of myself to work with.

METHODS FOR FINDING THE FROZEN ISSUES: I have used two main lines of questioning (Six Areas of My Life and Five Element Process) that provide useful answers to the issues that may have stopped the maturing process at that particular age. Sometimes I combine these two main methods of questioning. I recommend using GAME (Googleverse Access Method Express) or muscle testing to access information beyond brain thinking. Consequently, the issue can be discovered quickly by asking the following lines of questions using these methods, which cover the main areas of one's living. Here is the outline repeated from Pregame Decade 4.

The six areas of my life include the following: family, school, job/career, death/sickness, moving my location, and religion.
First line of questioning**: The Six Areas of My Life:**

7. Is it family? If yes, delve deeper by asking:
    a. Is it about parents: mother or father?
    b. Is it about siblings, brother or sister?
    c. Grandparents?
    d. Spouse?
    e. My children?
8. Is it about schooling? If yes, delve deeper.
    a. Teacher?
    b. Fellow students?
    c. A subject or grades in school?
9. Is it about job or career?
    a. Boss?
    b. Colleagues?
10. Is it about a death or a sickness?
11. Is it about moving my location?
12. Is it about religion?...

The second line of questioning uses the **FIVE ELEMENT PROCESS** that I have developed and their BATHWAVEs.
Step 1): Choose an element using my method: fire, earth metal, water, or wood.
Step 2): When using the GAME choose a single number from the bag that must be less than 8. Then choose that many BATHWAVES from the BATHWAVE bag.
Step 3): Access The element BATHWAVE charts which can be found in Decades 7-14. Or in the appendix.
Step 4): Do my replacings and write the present tense only at the end of the chapter.
In this work, I do not look for stories. I look for the **verb** or (**action**) and occasionally the **feeling** (**adverb**), or a combination of these two. I then use the mirror-call and Ace of BATHWAVE techniques.

Personal example of melting frozen ages:
I surprised myself when I muscle tested the maturity level of my inner child or "**creativity**". This aspect of me was frozen at a maturity level of 9 months old. This was right when I broke my infant left leg to the doctor's disbelief. I had run away from my mother when she wanted to bathe me, into her newly waxed kitchen floor and fell screaming in pain. Proven in an X-ray, I had a full leg cast sitting in a red wagon for my first birthday as I saw in a photo.

The only reason I write this story is to find what the verbs are of this story. The verbs that I chose are that I am **running** away from my self- nurturance **(mother)**, number one. Number two, I am also

angry and depressed at the restriction in the red wagon for many months while healing my broken leg. I am an extremely active personality.

Mirror-Call Method:
Speaking to my **mother** in a mirror. Angela, the child, says, "I am angry you are not more loveable and cuddly to me and I am running away from you".

When I look at figure 4.1 of what aspect of me my mother represents I find my "**self-nurturance**". I am her "**child**" in the conversation. The child aspect of me is my **creativity** dimension.

BATHWAVE TIME EXAMPLE FOR MELTING FROZEN AGE	
Part 1 BATHWAVES I Face and Embrace... in whole body	**Part 2 REPLACINGS** around my Midline with Hands, I Replace with Grace. I can, I will and I...
I face and embrace that my **self-nurturing** is not loving and cuddling **my creativity.**	My **self-nurturing** can, will, and is loving and cuddling my **feminine creativity**.
I face and embrace in my whole body and being that **my creativity** is angrily running away from **self-nurturing**.	I can, will and do focus freely and passionately **nurturing my creativity.**

Before I sleep I say my quantumcode and replacements with my eyes closed.

DAY 3
MELTING AGE-BERGS

MORNING PRACTICE: I begin my day by singing or saying my quantumcode and replacements with my eyes open.

How do I feel this morning? After maturing two aspects of myself do I notice any differences relating to myself or others? Does my body feel different? Reflect and write about it. Be aware of any differences in my interactions today.

Personal example of melting frozen ages:

I am now ready to muscle test this quantum replacement age starting at 1 year old where I left off, progressing to 5 years old then 15 years old and 20. Twenty years old tested no or weak. I go back to 15 years old and count by ones and get a no at 17 years old. I ask myself what stopped my self-nurturance at 16 to 17 years old. And the answer is that I was about to go to college and felt a lot of ***fear of failing***.

BATHWAVE TIME EXAMPLE FOR MELTING FROZEN AGE	
Part 1. BATHWAVES I Face and Embrace in my whole body	**Part 2. REPLACINGS** around my Midline with Hands, I Replace with Grace.
I face and embrace in my whole body and being that I am **afraid of failing.**	I can, will, and I am **confident of succeeding in whatever I do.**

EVENING PRACTICE:
I can continue melting the age-bergs process this evening. Use a method to determine which aspect to work on this evening and find the first age I am frozen at. Find the issue using the two lines of questioning: Six Areas of My Life and Five Elements Process as I begin melting my frozen age continue to the age of 200.

Before sleep listen to my recording and say my quantumcode and replacements with my eyes closed. Ask for a dream about one of my freed up aspects.

DAY 4
BRRR...

MORNING PRACTICE: Good morning! Say my quantumcode and replacements with my eyes open. I am halfway done melting the frozen ages of my six aspects. Is there a frozen age that surprised myself? Write about it and how I felt after BATHWAVING my replacement.

EVENING PRACTICE:
This evening I can complete my previous aspect if needed. Continue working through the process of melting my frozen ages. Use a method to determine one of my remaining aspects to unfreeze and repeat the process.

Before I sleep I say my quantumcode and replacements with my eyes closed.

DAY 5
I'M MELTING...A FROZEN AGE ASPECT OF MYSELF

MORNING PRACTICE: My quantumcode and replacements ARRRRR...Replendent when spoken with my eyes open. Do I feel like I have melted? Is my body more relaxed? Have I noticed any changes in my relationship with myself? Write about it.

EVENING PRACTICE:
Do I need to finish any work from this morning or evening? If not, I can continue working through the process of melting my frozen ages. Choose one of my remaining aspects.

Before I drift off to dream I say my quantumcode and replacements with my eyes closed. I can always ask for a dream about the new me.

DAY 6
REWRITING YOUR PAST IS A QUANTUM GIFT

MORNING PRACTICE: A beautiful way to start my morning is to say or sing my quantumcode and replacements with my eyes open. This evening I will work on my last aspect of myself. During the day be aware of this aspect of myself, do I notice this aspect in the relationships I encounter with others?

EVENING PRACTICE:
First I can finish any melting from the previous day? If already complete, I can continue with my last aspect of myself. Once finished I can recite my quantumcode and replacements with my eyes closed. Is there anything I desire my dreams to reveal to me, ask away?

DAY 7
NEW 'UNFROZEN' LIVING

MORNING PRACTICE: Say my quantumcode and replacements with my eyes open. Now that I have melted all aspects of myself; do I feel different this morning? Take a moment to reflect on this process, how did I find this process? What feelings or thoughts arose during this process? I may need to face and embrace these feelings or ideas.

EVENING PRACTICE:
This evening I can complete any melting of my ages I may need to complete. Once I have finished write down all quantum replacements in the present tense. These replacements are to be repeated for 21 days to 3 months after this process is complete. Before I sleep say my quantumcode along with old and new replacements with my eyes closed.

DAY 8
IN MY GIVING IS MY RECEIVING, IN MY RECEIVING IS MY GIVING

MORNING PRACTICE: I start my morning reciting my quantumcode and replacements with my eyes open.

Money is energy.
Money is a conversation.
Therefore energy (including $) is a conversation with my self and the universe.

Today I write all my conversations (thoughts, problems, and feelings about money). I notice the conversations I may have today with others involving money. Take notes and this evening I can replace the word 'money' with 'energy' and BATHWAVE them all.

EVENING PRACTICE:
This evening I face and embrace my conversations around money.

Part 1 BATHWAVES I Face and Embrace... in whole body	**Part 2 REPLACINGS** around my Midline with Hands, I Replace with Grace.
I embrace that money is scarce in my life. I need to not spend money on what I need or desire. I am ready to replace this. No waiting. These ideas are manifesting an undesirable impoverished living in me.	Energy can, will, and is abundant in my life. I can, will, and am using my energy fulfilling true desires (as seen as my abidingself). I can, will and do today what I creatively desire to accomplish.

DAY 9
SOMETIMES MONEY GETS A "BAD" RAP

Culturally there are some attitudes against money. Some cultures think money is bad, less money means more illumination, I am trapped in my socioeconomic status, or that scarcity is necessary. Cultural and personal beliefs, actions, thoughts, habits, words, attitudes, values, and emotions around money influence energy (money).

MORNING PRACTICE: I begin by saying my quantumcode and replacements with my eyes open. Notice any attitudes I may have against money and people who have money. Write about these attitudes.

EVENING PRACTICE:
This evening I face and embrace my conversations around money.

Part 1 BATHWAVES I Face and Embrace... in whole body	**Part 2 REPLACINGS** around my Midline with Hands, I Replace with Grace.
I face and embrace my attitudes against money or these attitudes will push money away.	I can, I will, and I do as I desire manifesting the energy and doing it as I truly desire.

Before bed I say my quantumcode and replacements with my eyes closed. Ask for a dream about my attitudes or conversations around energy/money.

DAY 10
EXCHANGES OF ENERGY MANIFEST WAVES (WOLF, M. (2008))

Vision it forward: Exchange is energy (money).

MORNING PRACTICE: After waking say my quantumcode and replacements with my eyes open. Create pictures of myself doing what I desire and receiving exchanges of energy ($).
Make a recording of my new living speaking in present tense using these pictures. For example: I have plenty of energy. I am traveling and share my triune wholeness with others. I have a partner who is joining me and/or supporting me in living my true desires. My children and friends are energetically fulfilling their own desires as they choose. Continue this with all the images I desire. Listen to this for as many days as it takes to become automatic.

EVENING PRACTICE:
Money is also a form of self-acknowledgement. Am I trying to prove myself and get acknowledgement for it? It's BATHWAVEing time; I face and embrace this. I replace it with: I acknowledge myself as valuable in everything I say and do (my relationships, my work, my dreams, my body, my creations, and my cooking). This expands the universe in priceless valuable waves.

Congratulations write down all quantum replacements in the present tense. Before I go to sleep for the night I recite my quantumcode and replacements with my eyes closed.

QUANTUM REPLACEMENTS FOR
MELTING FROZEN AGES & INVITING ENERGY/MONEY

I will repeat these present tense replacements for at least 21 days to
3 months. Morning eyes open and evening eyes closed.

My self-nurturance is loving and cuddling my feminine creativity.
I focus freely and passionately nurturing my creativity.
I am confident of succeeding in whatever I do.
Energy is abundant in my life.
I am using my energy fulfilling true desires (as seen as my abidingself).
I do today what I creatively desire to accomplish.
I do as I desire manifesting the energy doing it as I need.

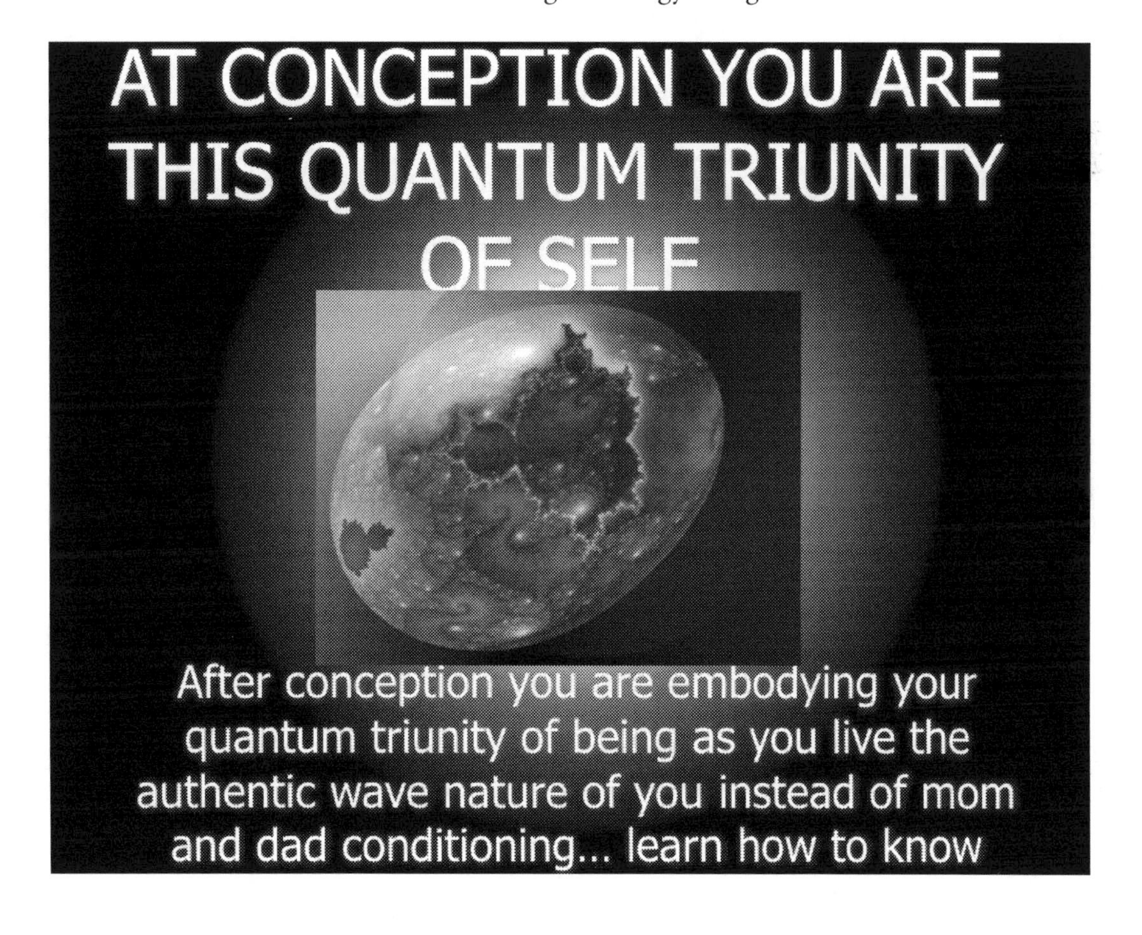

DECADE 16

MY ORIGINAL STORY IS MY ESSENCE AND INCIPIT*

DAY 1
WHAT IS MY ORIGINAL STORY OR ESSENCE

MORNING PRACTICE: After waking say my quantumcode and replacements with my eyes open.

My original story is the ESSENCE OF ME. It is the theme that is motivating my abidingself. I would say that it fuels the declaring of my quantum wave particular pattern.

This original story is so important and usually is perceived later in this work after I have been aware of the nature of my quantumcode for many months. I will begin to be aware of it. I may see this pattern early in living.

My original story, also called my 'incipit', manifests immediately or is resisted early in my life. It usually can be seen as a repeating pattern. If I resist it, then I will perceive a negative repeating pattern that is the mask or filter, which has been covering and hiding or clouding the expression of my original theme.

This incipit motivates the formation of my unique abiding quantumcode. This pattern that I have been living a long time can be identified through my relationship with my environment as a personal being. I can't remove the filter. I can only replace it when I have gained enough awareness of that. Then I can literally embrace living the original story.

As I embrace fully the original story then the abidingself is more easily able to manifest its actions. The incipit is recorded in the universe.

Clients say they feel a new lightness never before experienced and that a weight was lifted from them. This will last as long as they remember to live their original story.

Do automatic writing about what I loved most about childhood. Make it up if I have no idea, like a story or novel. I can even write it about someone else's childhood. Just write whatever comes to mind about childhood.

EVENING PRACTICE:

EXAMPLE OF ORIGINAL STORY DISCOVERY FROM A DAILY EXPERIENCE WITH A COACH'S ASSISTANCE WRITTEN BY CLIENT

"After four months of working closely with Dr. Angela Longo, we had developed a close, personal friendship. I was assisting her with editing this workbook while receiving training to become a Quantum Living Coach. One day I had an experience that was triggered while caring for a toddler. This trigger led us to discover my incipit, my essence, my original story.

I heard the stirs of a toddler waking from sleep. Approaching the room I was prepared for some tears to be shed knowing that when she fell asleep her mother was here. She would most likely expect her mother, not an unfamiliar adult. I knew I would need to give the child some time to adjust to this surprise change. I had been caring for children for 20+ years so this wasn't my first rodeo.

The tears were continuous during our time together subsiding for moments to play and interact with a sudden outburst of tears and a cry out for mommy! We tried walking, the park, eating, singing, yet the repetitive pattern of tears and an outcry for mommy continued. This was a new experience for me as in my past the child would shed some tears and after a few moments we were all able to move on.

Knowing about mirror calls I was able to identify some patterns to replace with my own mommy (my self-nurturance). The replacements resulted in calmness to follow for a miraculous moment in time, yet didn't stick for the duration. Once her mom arrived, all was right in the world for the child.

In reflecting this experience with Dr. Angela she suggested that I spend more time with this experience and she coached me through it.

I am 8 months old crying for my mommy, yet my mommy isn't here and she isn't coming. I feel alone and sad so I cry more. There's a pain in my gut and my jaw tenses. Why am I not being heard, why is no one attending to me? I am an infant, a child, I am important, acknowledge me for I am communicating with you, I need you.

What my creativity receives and understands is that adults do not value a child, they do not think I am intelligent or important, yet I know that I am, I am intelligent, and what I say and do is of value! I am intelligent, and what I say or do is of value! I am important, I am expanding, I am expanding the Universe, yet I still feel this sadness of alone.

My presence is at inception, I am here and it feels cold and not loving. I embrace this coldness, I love this coldness, this coldness is just my judgment AND MY JUDGING, a lack of connecting, a lack of acknowledging, a lack of nourishing, a lack of unconditional non-judging. I embrace this conditional judging at the egg and sperm level. There is no room for that any more, "I can, I will, and I am living my unconditional love which is self-unjudging (that's what love is)". My original essence, my original story is revealed, "I am the essence of warming nonjudging interacting (play) as Kelly Desires! Whoopee! This is ME, my essence, my original story. I have now become the Quantum Playgirl!" (Wedin, K., 2018)

Ask for a dream of my personal incipit. Say my quantum code and replacements out loud with eyes closed.

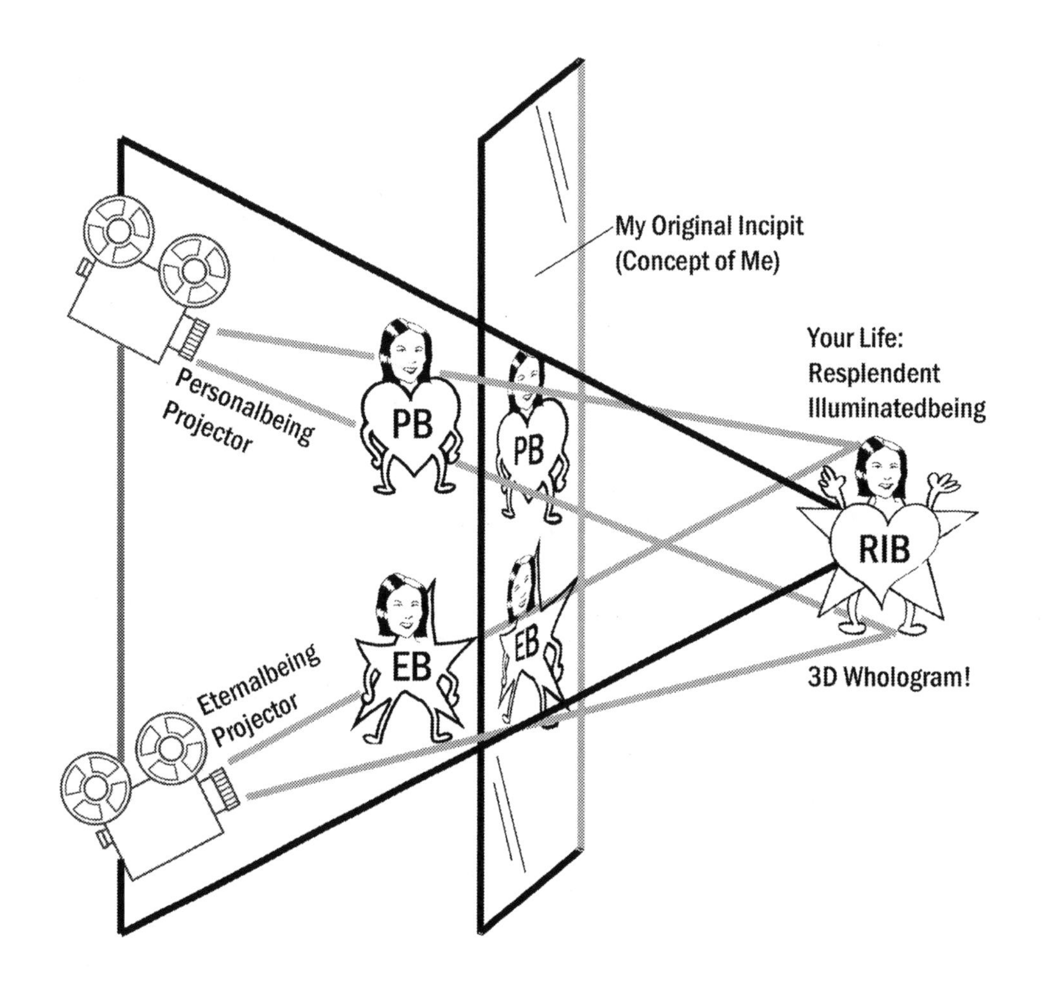

DAY 2
EXAMPLE OF INCIPIT DISCOVERED FROM CHILDHOOD DAILY HORRORS
MANIFESTING IN BODILY SYMPTOMS

MORNING PRACTICE: Say my quantum code and replacements out loud with eyes open.
Do some automatic writing about what or who bothered me from early childhood. If no idea, write about the diseases that bothered me in or from childhood.

EVENING PRACTICE:
This process is so profound and so powerful that usually it takes months into this work to discover and sometimes years to embody.

I would like to cite an example of an incipit discovery through horrible childhood resistance.
When I met a woman named Bella in her late 30's, she was suffering from horrific headaches. She would not be able to help herself at all during these headaches. No medication would touch these

headaches that were more horrible than migraines. As she unveiled patterns from childhood the stories were equally as horrible. She would cry and rock forward and backward as children do when they are stuck in a pattern.

The story basically was that she and her sister would witness her father yelling at the mother and then throw things at the walls. It is possible she said that he would throw her mother too. I began to picture objects like chairs that were thrown, how scary. Poor Bella would sit up at night as a child and out of frozen fear just cry and be powerless because there was nothing she could do to change her father. He never rose up against her, however he did against her sister. This was the background and nobody had been able to touch this in her. She suffered terribly.

After introducing her to the work her headaches had much improved from the original state, but still, she suffered an infection in her urinary tract of E. coli that was very painful. She said this symptom was a recurring pattern and that began to show me that she was holding onto the 'shitty' patterns that were "pissing her off. It was time to discover Bella's original story.

This is what came recently of the culmination of many many dreams. She admitted she would get very very angry at women who she judged are not very intelligent. She would lose her patience and of course we can see the pattern here with her mother.

She also realized in herself that she could not reawaken her own femininity. That she really loved her masculinity, but did not want to bring out her own femininity. So as she embraced this pattern, she got the real image of her father and mother. Of course, the father is self-guidance. Her self-guidance (masculine) was bored (angry) and out of ignorance blamed and judged the stupidity of the feminine. To save her own life as she had to in that moment of fear, she resorted to just hating the fact that the mother appeared very ignorant. Her intelligence befriending the father, saved her, she emulated masculinity throughout her life.

In her first BATHWAVE she embraced that her creativity (child) disliked her self-nurturance (mother), which is the mirror-call way of saying that as a child she disliked her mother's lack of power and lack of intelligence in that moment. Embracing that in her body and replacing it with: my creativity understands and appreciates my self-nurturance.

After that in her second BATHWAVE we embraced that her self-guidance was bored, violently angry and closed. She replaced it with her self-guidance was calm, open and enthusiastic with the brilliance and illuminated self-nurturance.

I asked her to picture what the original story self would look like. She said it looks like a little girl with big blue eyes that sparkle like mine. I said, "What are the eyes saying?". She said, "The eyes are saying illumination. They are bright and beautiful.".

Of course, it was an image of herself the last time she remembered feeling her original story. Filters of fear and boredom (anger) with powerlessness plagued her living. She was able to reclaim her original story, which you can hear behind the name "Reveille" (meaning "awakening" in French).** This was the name she gave her abidingself, which is the source of "the sacredness (brightness) of the meaning I give to living." Her particular meaning she gives to life is the opening (femininity) of the essence of the sacredness of the universe and of living. This brought tears of joy to her eyes with this original story.

**Bella's quantum wave pattern: Reveille offers looking into eyes to show the opening of the space of the illuminatedbeing, while engaging directly as my particular universal illuminatedbeing.

Of course, this is a picture in my mind. The incipit then is the motivation for the formation and living of the eternalself. Probably, the second most important process in this work, the first being the actual reading of the pattern of the abidingself. That existence and the nature of that helps us to bring forth the strength and the power to both identify and live this original story. That is why it has taken a minimum of six months to identify typically.

Ask for another dream of my resistance or mask to living my original story, my incipit.
Say my quantum code and replacements out loud with eyes closed.

DAY 3
YOUR ORIGINAL STORY IS THE MEDIUM CALLED "SPACE" THAT MY QUANTUM WAVE MANIFESTS IN

MORNING PRACTICE: Say my quantum code and replacements out loud with eyes open.

For all us nerds out there, the quantum world picture of what my original story would be is the medium called "space". Space is a medium in order to exist. I would hear no music without air. Air creates the sound waves that travel to my ear to be heard and perceived. Likewise my abiding self manifests in the medium of my essence, which is my incipit or in my original story. The quantum wave is carried in the medium of space, which is the big frontier. We don't really know what space is right now. In the quantum world because every human's quantum wave is unique, it is the essence of me; it is my medium. That is unique for me and is chosen just at incarnation which makes sense to me. These themes are in the universal quantum memory otherwise known as the quantum entanglement records or in India Akashic records.

So this incipit is the big space of myself, which is the medium that I choose to manifest my abiding self in. My quantum wave pattern manifests in and is relative to the original story and that's why it's such a gift.

Maginate (focus on the idea) and feel myself reaching out into space in a meaningful way today.

EVENING PRACTICE:
A professor of quantum physics named Wheeler said "Matter tells space what it is. Space tells matter how to behave."

I would paraphrase that to say my personalself body/brain tells my original story what it is. The original story tells personalself body/brain how to behave, as my quantum wave pattern, which is "what I do and how I do it". The personalself in living its incipit formulates its quantum wave pattern as it interacts with the environment and people, as I desire.

So it is very creative living with this awareness and it's all my choices. I do make the choices. I may resist the original story as I will give an example. Many people resist their original story and that's why their childhood is filled with illness or fear or anger/abuse or other forms of reactivity like allergies, etc. . Resistance can be reactivity; reactivity can be resistance.

Once I perceive my original story then I am sailing as an abidingself/personal self/resplendent illuminatedself to the world of conditions of expansion and resplendence. This is what I can live in this moment.

This knowing is not easy to perceive by myself. Processes I am aware of:

1) I can look at childhood and summarize the theme that I am either living or resisting or both.

2) I can look for words that express the actions that made my eyes sparkle when I did it as a child. Or what people said about me as a baby. It will be actions with a little description possible.

3) Do this practice tonight:
If I had all the money in the world and no responsibilities what would I love to do and then do automatic writing. Ask for dreams about it. Write down dreams and replace what I do not desire. A few days after that, I can read all that and look at what theme I really am doing in these processes. That might help me describe in a few words my original story, my incipit.

4) I can also ask for a dream revealing my original story. Ask for guidance if I don't understand the dream.
Say my quantumcode and replacements out loud with eyes closed.

DAY 4
AWARENESS OF MY ORIGINAL STORY BENEFITS ALL

MORNING PRACTICE: Say my quantum code and replacements out loud with eyes open.

Did the dreams I had help me describe in a few words my original story, my incipit?
Ask for a daily event that will help reveal my incipit today. Write it down even if I don't understand how yet.

EVENING PRACTICE:
It is very important to understand that when I do these processes, I not only benefit myself which of course is most important, I also benefit and re-write the past. I benefit those who have also been

part of this past with me. Third, I benefit everyone in the universe who in anyway shape or form in entanglement with this resistant pattern that I am replacing. Most importantly, in living my incipit, I am resurrecting, giving rebirth to, originating again, freeing, and opening to living my resplendence.

Make a list of everyone that I know who would benefit by my awareness living my original story. Say my quantum code and replacements out loud with eyes closed.

DAY 5
DISCOVERING MY EMPRISE

MORNING PRACTICE: Say my quantum code and replacements out loud.

Now that I know my abidingself and about my original story of "how I do everything in my life", there is a third piece of the quantum picture that is useful.

It's the global image of "what is the big picture that I would like to contribute, that I cannot complete in one lifetime"? Roger Cotting, my philosopher mentor calls this my "emprise".

I might read all of my six accomplishments quickly and ask myself if they have an overall theme or direction that is bigger than one life's accomplishment? Write it down in my notebook.

It's where my thoughts go when I daydream or think. I can ask a close friend or family member what they think it might be. The point is to pick a big dream of mine and not worry about being correct. I can change or expand it as I go thru this process.

I might ask for a dream revealing my emprise. Keep a page in my notebook with my ideas of what it might be. It is good to know later in my life.

An example might be that Angela's emprise is to apply the whole quantum understanding of the universe to living and share that with everyone that I can interest in it. That might just take longer than my lifetime.

EVENING PRACTICE:

Say my quantum replacements twice out loud before sleep. Once with my eyes open and once with my eyes closed. Say the sentence I wrote of my eternalself by heart, add my emprise if I came up with one.

WHAT IS RESPLENDENT RELATIONSHIP?
THE REAL R&R

DAY 6
MY RESPLENDENT TEAM

MORNING PRACTICE: Say my quantumcode plus replacements eyes open.

What is this resplendent relationship? Well the first is within my own selves, since I am a team. There are three selves to begin with: my personalself and my abidingself. When those two relate I have an emergent resplendent illuminatedself witnessing my personal and abidingselves.

I know my personalself as my body and brain. My personalself becomes aware of my eternalself really as vast as the universe, then RIB begins to relate to those aspects of myselves in everything I do: my dreams, my symptoms, my daily events, relationships and creations. These relationship interactions are the embodiment into my 3 beings: the 'exchanges' between Angela, personalbeing, and my emergent quantum wave patternbeing called 'Communiona' are being witnessed by my emergent resplendent illuminatedbeing. Hence, the awareness of 6 of me.

Diagram of the Virtual TriUnity
in Communion with Adorned TriUnity

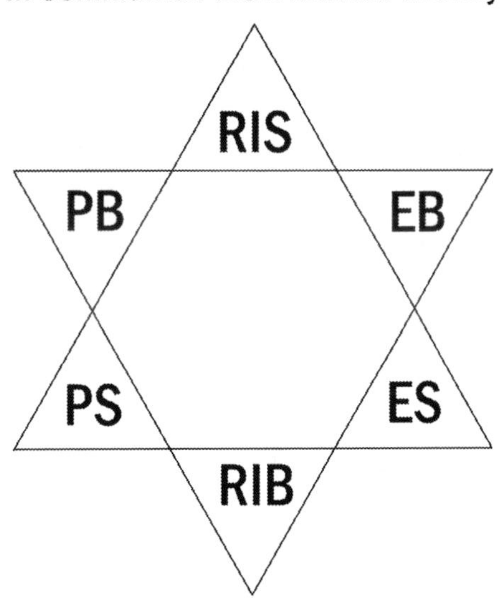

Basic Building Block of Sacred Geometry's
Flower of Life

Be aware of witnessing my interactions today with myself or others as they mirror my interactions. Spend time at meals looking to notice any self- witnessing or conversation in the moments prior. Anytime I say anything to myself, even judging, I am witnessing myself first!

EVENING PRACTICE:

EXAMPLE OF DREAMTIME RESPLENDENT RELATIONSHIP

When I awaken, my three beings can be aware of the possibility of living new options revealed in understanding the dreams. Mostly, I have been ignoring these valuable dream communications from my resplendentselves.

When I am in the awareness of my resplendentbeing, dreams are a very good example of this relationship. In the dream the quantum wave patternself is relating options to my personalbeing, Angela.

As I look at the first image and action of the dream, holographically, I can see what the whole dream is telling me. I now have options and then becoming aware of these options I am a resplendentself living. So now we have those three aspects of myselves alive and well.

Everyone has these three aspects of selves. When I embody it in my living they become three beings. As was explained, there are really six of me.

Ask for a dream showing the six of me. It is ok if I can't remember it.
Say my quantumcode plus replacements with eyes closed.

DAY 7
RESPLENDENT HUMAN RELATIONSHIP

MORNING PRACTICE: Say my quantumcode plus replacements with my eyes open.

When I meet up with another human being, they also have these six companions that make a relationship between us 12. So what does that look like? It is quite exciting.

When I have a dream, sometimes I can be too close to the dream to see the options. When someone else has some space and distance from me they have an ability to reflect to me what options I am giving myself. They are supporting my resplendentbeing at this time by stepping into that position for me.

The second gift of the resplendent relationship is the mirror-calling of myselves. How can I truly know myself except to see myselves more clearly in a mirror. Human beings are those mirrors calling us to our true nature. When I can honestly and openly see in their behaviors, my responses, I begin to see my patterns. This vision of myselves gives more material to replace the patterns with and enjoy it at the same time, or not.

This really begins to look like a play, when six of me relate to six of you, awarefully then the play is amazing. The theme is self illumination. It is very creative and very expansive. The foundation of the universe is built on exchange. I have been describing personal exchanges that I do, between my selves with my dreams, my daily events, my symptoms, creations and my relationships and that's just the beginning. When I relate to another I have this creative ability to exchange the options they present me with, for me to replace.

Choose to perceive every communication I receive today from everyone as a communication from my abidingbeing by showing gratitude with a sincere smile or other such response. This smile would be coming from my whole triune resplendentbeing.

If I have a different reaction to a person or incident write it down to replace later.

EVENING PRACTICE:

UPLOADING GOOGLEVERSE FOR MY EMERGENT MIRACLES

Resplendent relationship gives us a great ability to access googleverse. My emergent miracles are instant manifestations of the interaction of our personalself and abidingself. They are life partners. So as I respond in relating, I can turn my tantrums into tantra. All I have to do is replace the 'rum' with a 'ra'. Rum represents the numb drunk self. As an example, when I drink excessive rum alcohol regularly, I can numb what I really want to be aware of and this is not where I want to go to have a resplendent relationship. I want to be awake to myselves in each moment. This isn't always easy but as soon as I see the option because of a pattern, I can replace it instantly and this is a gift of the quantum world understanding. The benefits far outweigh any discomfort.

Ask for a dream revealing my tantrums to tantra!

Say my quantumcode plus replacements eyes closed.

DAY 8
THE NEW ABIDINGBEING IN MY HUMAN RELATIONSHIPS

MORNING PRACTICE: Recite my quantumcode plus replacements with eyes wide open.

When I have a desirable relationship, sometimes I trap my selves into thinking I should keep the relationship the way it is because it's so good. This is a huge mistake. This emergent miracle requires new information, which we call in-for-motion, added to the relationship in each moment. I bring that to myselves through entanglement.

Entanglement is the quantum process that gives us access to all the information of the universe that has ever been. I bring that information as I desire into my life and then in conjunction with another that just makes our information exponentially expand. There is the information of mine and then the information of yours, which I bring together and start manifesting new emergent miracles between us.

So there's like a new life between us emergent of our relationship, a new quantumcode self. A new pattern comes together if both abidingselves are put together. It happens automatically as we interact. I can add the verbs together and play with the words and I see what picture I get. Then I have a new eternalself for our resplendentselves to relate to. This is expanding.

**Illustration of this entangled relationship emerging miracles.

EVENING PRACTICE:

MY RESPLENDENTBEING, WHETHER INDIVIDUAL OR IN RELATIONSHIP WITH OTHERS, MANIFESTS 'CONDITIONS OF EXPANSION'

After I am living this awareness for a while, my resplendentbeing relates to my abidingbeing and I begin manifesting conditions or states of expansion. I see the big picture expanding my living. This is the exciting gift resplendentbeings give to each other and offer the universe. I choose to call these: sacred conditions or states of expansion.

It's about how I live, so I don't have to be in a relationship with another human being to do that. I can do it right now. I can do this with my beings: with my own resplendentbeing relating to my own eternal abidingbeing manifesting these sacred conditions of expansion. Then personalbeing, Angela, becomes aware of these new conditions of expansion and I begin again, yet with universal expanded awareness! This is resplendent living.*

Ask for dreams revealing my conditions of expansion.

Say my quantumcode plus replacements eyes closed.

*All of this is only seen in hindsight. Don't look for it or try to prove it.

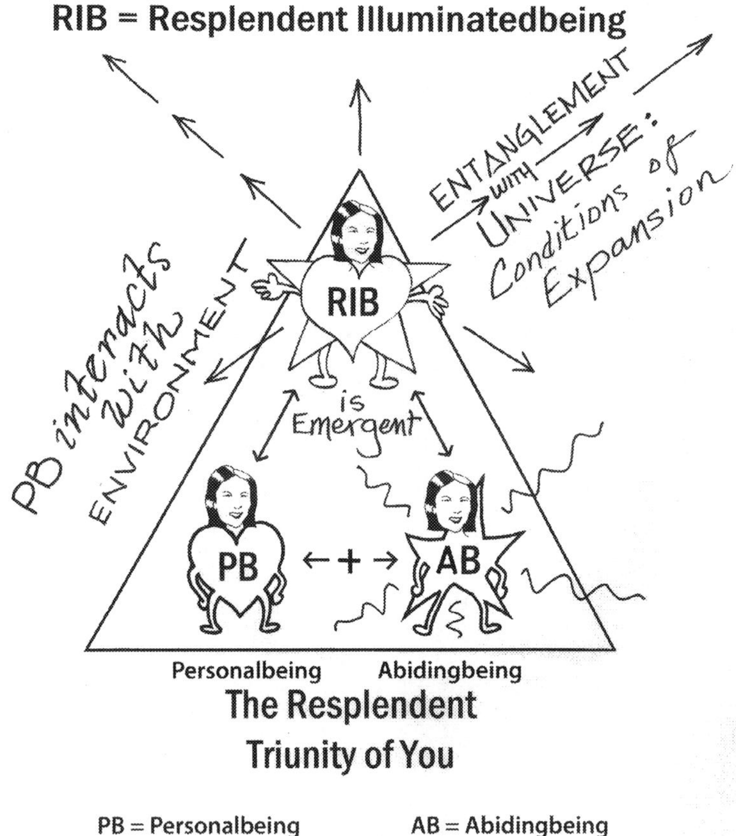

DAY 9
QUANTUM WAVE LIVING IS AN IMPORTANT CONVERSATION BETWEEN MYSELVES THIS MOMENT

MORNING PRACTICE: Say my quantumcode plus replacements eyes open.

GLOW TEAM!!!

I feel now that I am able to complete this workbook. I'm so amused, yet so grateful, because now I know why it's taken me so long to do this. I am sharing all this because I think it will be of benefit. I also understand why I need to write my books in conversation with my editor and/or mentors. Because it is teamwork and I need practice GLOWING as a QUANTUM TEAM!

I also realize that I need to go out and talk around the world live. YouTube never captures the alive me. I've studied with Lisa Nichols and now I know why the greatest orators speak live around the world. It's the conversation that is important.

I have been realizing that this is a principle of the quantum world: entanglement. Emergent miracles require interaction with the universe: this moment, and people are all universes, including me! I can interact with people as universes and it creates a whole new creative expansion of living in this moment.

What would I say to the universe? On TV or Oprah, if I knew that other civilizations were listening to me, what would I say? That really expands my creativity. Like the movie *Arrival*- everything becomes possible. We can value everyone's form of communication no matter the style. It is a little bit like the expansion of the money system to cryptocurrency. Everything becomes communication. Even money becomes a form of communication. Every form of creativity and expression including work is a form of communication. **My living and the mundane actions of my life become important communications to the universe. Because it's not what I do that's important, it's how I do it!**

This is what quantum communication is really making me aware of. That every single breath I take communicates to the whole universe. So my life becomes infinitely important and meaningful. What a relief!

Can I interact with a person today as a universe and maginate a whole new creative expansion of living in this moment?
Focus on my communications today in this image.

EVENING PRACTICE:

"INTERACTIONS GIVE ME NEW OPTIONS FOR EXCHANGE;

EXCHANGE IS THE FOUNDATION OF THE UNIVERSE"**

And the purpose of all interaction is exchange. Exchange was one of Roger B. Cotting's favorite words, since he said it was the foundation of the universe. (Cotting, R. B., 2018)

So there are no mistakes; there are only exchanges. If I do not like what I just did, exchange it. With this awareness of informotion, I am able to exchange instantly. I know that biochemically it works even with my parent's genes/DNA.

That's what I have trouble doing, exchanging!? I keep believing that I could lose something; which is never true in the quantum world. When someone or something leaves me, it is revealing that that aspect of me is leaving myself. **So I can exchange that in myself with commitment to myself or self-perseverance.**

Am I in grief or anger or fear about 'losing' someone or something? What aspect of myself am I leaving?
Whatever aspect of me they are mirroring, am I willing to commit to that aspect of myself? If I cannot figure it out, just embrace that I leave all of me. Finish the BATHWAVE this moment and add the replacement to my list.

Say my quantumcode plus replacements with my eyes closed.

DAY 10
RESPLENDENT EMBODIMENT IS HELPFUL FOR RELATIONSHIPS AND IS NEW THIS MOMENT

MORNING PRACTICE: Say my quantumcode plus replacements with my eyes open.

This idea of embodiment is so powerful and so subtle.

The body needs to be respected and honored because it is the manifestation of the whole universe. This is the quantum point of view. The body is quantum waves. The universe is quantum waves. All matter is quantum waves.

Remember, I am outgoing quantum wave as big as the universe bouncing back as my incoming quantum wave, which spins and it matters!
Next pulsation my particle disappears and my new particle replaces it.
These ideas are consistent with the work of Wolff, M. (2008). Please read his book Schrodinger's Universe if you want to learn more quantum physics.

My quantum waves manifest particles for me as fast as the speed of light. So why worry about what I have done? It doesn't make sense since it will be new again by the time I worry about it, so if I don't like it why can't I just do something different? ***When I am a creature of habit the same old, same mold is the real issue.***

Helping each other to enjoy something new is really where it's at. *Do something new; acquire new informotion! Even when I acquire only one new word or one new page or one new chapter or one new movement or one new friend, or one new creation, there is an emergent miracle for YOU. Just do it- don't think about it!*

To share something new, I ask my question, and the universe responds with my immediate answer. Answers are in the quantum memory, yet I do not know how to receive and understand it. The answers are in my body, my relationships, my dreams, my creations, everywhere, in every moment, and in everything... In the whole universe for quantum's sake!

That's the promise of entanglement: total access to the whole universe of information anytime. My answers are as deep as my questions. Some of my questions of patterns may go back into my past transforming my present. Time, as we define it in language, disappears. Greene, B. (2004)

QUANTUM WAVE LIVING IS SO 'REFRESHING'

This is so refreshing that I can have discussions with quantum memory whenever I desire. I am very emotional about this awareness; right now I am choking up and coughing. I am having many coughing orgasms, which we lovingly abbreviate as cocos!! Meaning I am replacing my old 'emotions' that don't serve me anymore which are expressed as 'mucus' in my lungs.

MY LIVING IS NOW 'REFRESHED' LIKE A WEBPAGE MADE NEW.

Do 'NEW' TODAY for at least one hour of my day then SHARE IT with someone. Everyday do BEING AND ACTING 'new' for another hour till I feel new all the time.

EVENING PRACTICE:

DISEMBODYING IS THE REAL ISSUE, NOT THE TECHNOLOGY OR THINKING

Using technology may be DISEMBODYING for us such as mobile phones, smart phones the whole movement toward artificial intelligence. Technology may distract me from awareness, which is in fact the next stage in evolution.* **Disembodying is the real issue not the technology.** Technology can be one of the addictive distractions from the awareness of my illuminatedself. It may be used as an attitude against authentic living. I will find a distraction if I want one, there are millions. Understanding the ultimate 'technology' or methods of quantum wave of living has totally enabled me to appreciate my body as a new manifestation in this moment. I no longer resent my body or living as something I have to carry around like baggage.

Thinking is not usually the issue; it is what we are thinking, which is mostly judging. Judging the body, life, or the world is usually the issue with disembodying.

DISTRACTIONS VS. MIRROR-CALLS

Anything that takes me away from enlivening my illuminatedbeing naturally is just my distraction. So I am very aware when I find myself distracting myself from authenticity.

STRESSED SPELLED BACKWARDS IS DESSERTS! SO STRESSED DISAPPEARS AS A WORD

We are so interested in having that house, long life, health, well-being, exercise, and expensive supplements. I believe that this may buy time, yet I'm realizing now that the new quantum illuminatedself awareness is the most important aspect for me.

All the wonderful treatments, supplements, medications, and exercising are all supportive, yet I find this new awareness the priority for me. I do not wish to underestimate the value of good relief as band aids.

Yet stressed disappears as a word. Or better yet, as I said in my last book, stressed spelled backwards is desserts! So everything becomes desserts when I see it quantum style, as a mirror-call.
Ask for a dream that will show me the desserts I am living!

As Einstein wrote:
"There are only two ways to live:
As though nothing is a miracle.
Or as though everything is a miracle!"

May you now see how everything you are living is a mirror-call (miracle) of your resplendentbeing!

GLOSSARY

This glossary is here to clarify one's understanding of what we talk about and introduces new words and descriptions. Reading these words helps me understand this workbook. Sometimes the words appear twice and are described in a different style.

Abiding self: The abiding self is the environment I create within which I personally live eternally.
My actions I have done that I am proud of. Action of being proud of.
My image of the universe is my image of my vast abiding self.
Vast abiding self eternally (VASE): is the action of how I do everything I am proud of.
If I am not focused on VASE then I don't have a consistent and reliable universe to be alive in.
My vast abiding self eternally is akin to the pattern of informotion that repeats itself like a quantum wave, which fills the universe instantly. It is so vast to grasp.

Action: Action is the only thing that the universe responds to.

Ace of BATHWAVEing is a simple technique to assist in the transformation of my beliefs, actions, thoughts, habits, words, attitudes, values, and emotions, which consists of 2 parts: face and embrace, then replace with my grace known as my true desires.

Aether: Aether is the space within the universe. Space within the universe is quantum memory. Aether and quantum memory are exactly the same. These are space, therefore space is not empty.
Physicists say space is full of quantum fluctuations.

BATHWAVESs is the acronym for my beliefs, actions, thoughts, habits, words, attitudes, values, and emotions, which are my main informotion manifesting my reality.

Bent time: Einstein theory of bending time which is memory through action creates a new memory. Einstein's theory of speed of light is correct in space. But not the same in memory. 10/6 (Cotting, R. B., 2018)

Center of the Universe: Every point of the universe is the center of the universe, since everything moves away from everything else at the same speed. They collapse to one point which I derive from my list of accomplishments as the encode decode key of my abidingself. This would be the center of my quantum wave.

Chi or qi or ki is the Chinese and oriental word for energy. They describe many kinds of chi in the body and universe.

Conditions of Expansion are the interactions of my aware witness called illuminatedbeing with my abidingbeing or that which I am proud of doing. These states of universal awareness are the expanding unique doings I naturally am expressing as my quantum wave pattern. Scared melts into sacred with this expansion.

Context is defined as being consistent to a single aspect.

Data: Data is like the alphabet, which says nothing. Words are information.

Data and Information: These are the constituents of space and are free and recorded in the universe from which we can draw and extract freely.

Descriptionary is like a dictionary except it uses descriptions of words instead of definitions. This glossary is a descriptionary.

Direct perception is focusing, is accessing, initiating, and originating whatever we choose to focus on... **direct perception... Directing quantum memory...**
Direct perception is like dreaming our dreams. When we ask for a memory/entanglement this is called **"direct perception" or "magination".**
Direct perception means to live what I truly desire without trying to cause it.

Emergent means coming from nothing, not caused from any of the constituents.
The properties of water are emergent from hydrogen and oxygen. Neither hydrogen nor oxygen have any of the properties like water.
The random coalescence of desired constituents, which gives a new and different consequence beyond our own imagination.
Every action or extraction of constituents gives an emergent miracle instantly. When constituents are extracted from the universe, they come together randomly to give us more than we imagined to replace the old.

Emergent process: An example of an emergent process is the coalescence of hydrogen, which is flammable, and oxygen, which is another gas that is flammable coming together randomly to form water, which does not resemble its constituents at all. Water is new and different and replaces the original constituents. This is the emergent process, which is the functioning or the basic action in the universe.
The misunderstanding called cause and effect is really a distortion of emergent.

Entinglement or envolvement: This is the new word assigned to the old physics term, "entanglement", which basically means that a quantum particle wave is simultaneously moving as any other particle it has met in the same but in a mirrored way instantly. Because all space within the universe is memory then it functions like a solid...this is what accounts for "entanglement". "So when you push here and there the whole moves simultaneously. When you pull on one piece of a solid the whole solid moves." (Cotting, R. B., 2018).

Emprise is the big picture of how you are living your personalself from which the vast abidingself eternally is emergent.
"**Emprise**: an enterprise or undertaking that cannot be completed in one lifetime! ...that you continue!" (Cotting, R.B., 2018)
Your emprise purpose is to keep you focused on continuing, since you cannot complete it in one lifetime.
Focusing on your emprise, which is how you are living your vast abiding self eternally, gives you a consistent, reliable, continuing existence though discontinuous. It's like playing a G note then stopping over and over again...consistently G note though discontinuous in playing.

Force is a quantitative description of an object being moved or acting upon.

Energy: is the coalescence of information.

Environment: Environment is memory. memory is recorded in **aether.**

(Quantum) Entanglement theory says that quantum 'particles' will mirror each others behaviour instantly. 'Particles' are understood presently as spinning centers of quantum waves. Picturing two waves connected through 'entanglement', it is easier for me to imagine how the 'particle' in their center might behave in a mirrored fashion.
Einstein called entanglement "Spooky action at a distance" since it challenges the very core of science's belief in cause and effect.

Entanglement, 'entinglement' or envolvement: This is the new word assigned to the old physics term, "entanglement", which basically means that a quantum particle wave is simultaneously moving as any other particle it has met in the same but in a mirrored way instantly. Because all space within the universe is memory then it functions like a solid...this is what accounts for "entanglement". "So when you push here and there the whole moves simultaneously. When you pull on one piece of a solid the whole solid moves." (Cotting, R. B., 2018).

Entanglement manifests quantum memory.
It is the quantum memory between things that we also called gravity. So entanglement is gravity... This is what scientists are discovering.
The relationship between things expresses **time or space or quantum memory and entanglement.**
Informotion is active relationship.
Everything is a relationship in the quantum world.* Capra, F. (1969)
The **quantum wave** is made up of informotion, therefore, is a relationship. Entanglement makes universal relationship a reality and easier to understand.

Essence: Your essence or original story is prior to your abidingself. Your abiding self is the active manifest essence of how you live you. You are not living your abidingself. Your abiding self lives you.

Living as my personal self is the misunderstood way around, and if I did I would have no image of my abiding self. The abiding self has the image of my personal self. (The quantum wave spins the appearance of particle into existence.)
Essences can also be a tool, story, or process of reminding myself with writings, questions or oil fragrances, how to select information to support and live the life that I desire in a playful way.

Eternal is described as recurrring over and over again.

Eternalbeing: My embodied quantum wave-like pattern = my quantumcodebeing = my abidingbeing (meaning: commits to my personalbeing) **= my eternalbeing** (meaning: my recurring pattern throughout my living and throughout the universe recorded in quantum memory forever)**.** These words may be used interchangeably throughout this book to help me remember the qualities of this forever companion.

Eternalself: My virtual **quantum wave-like pattern** = **my quantumcode self** = **my abidingself** (meaning: commits to my personalself) = **my eternalself** (meaning: my recurring pattern throughout my living and throughout the universe recorded in quantum memory forever). These words may be used interchangeably throughout this book to help me remember the qualities of this forever companion.

Five Element Theory is an oriental energetic construct describing the nature of relationships between body organs and their physiology, psychology and quantum wave nature. Patterns as described in the I Ching, I believe, are the beginning of an understanding of our 'informotion, as is described in this book.
I plan one day to complete what Roger B. Cotting began in rewriting the I Ching from the quantum point of view as a tool of Resplendence.

Heartwaving was the original name given to the process of replacing BATHWAVES.

Imaginative minding is that function I do to move stuff around in an emergent way.

Incipit is the original story I choose at incarnation. It could be called the essence of me from which my abidingself manifests by my personalself interacting with the environment.

Information is like the words made from the alphabet.

Informotion is the motion that is the reason for the form or actions of the data or us.
In reference to a quantum wave, it does not require movement as it is a standing wave.

Light itself does not travel through space...instead light illuminates space which passes it on in a direct way....then light may illuminate "quantum memory" or informotion....this may help physicists determine what "quantum memory" is and therefore what space and "dark matter" and "informotion" are comprised of...

Living is a **process** not a conclusion.
Living is a journey not a destination.
Promise is a way of living, not a thing you do.
Your body is your continuing **soul**...

Love letters are the mirrored manifestations of our quantum wave patterns of informotion which give us options to add new informotion as we desire..

Magination means to focus on and see what I truly desire...

Meaning: Meaning means we get to live any meaning we desire with a reactive universe. Since the universe is only reacting to what you do, it's living your life as you live, right beside you.

Memories are defined as that which is stored in our brains and used as 'pieces' of data/ informotion of the past.
(Particularized cells function through memories of the past.)

(Quantum) Memory is an event that is stored and used as a whole event of information. All past occurences stored in the universe on a quantum level possibly in the informotion of dark "matter". Quantum Memory is space and time... You can directly perceive quantum memory.
Everything material is quantum memory studied as the science of psychometry.
Then the question arises is quantum memory a quantum wave fabric of the universe?
You manifest the universe from "quantum memory"...
"**Memory** is stored information as a whole event in Aether. Memory is connected by memory. Memory is instant because all space is memory. Memory functions like solid. Push here and there it moves simultaneously." Cotting, R. B., (2012).
Sounds like entanglement. Longo, A.

Mind: Data and information of the universe.

Minding: Minding is moving data and information around or reordering it around.

Mirror–calls are the phenomena appearing in our living that we respond to in anyway shape or form. These are considered mirrors reflecting aspects of ourselves as gifts of transformation. These may be encouraged by the process of entanglement which we have replaced with the user friendly word 'entinglement'.
Saying it a different way:
Mirror-call is a quantum descriptionary word for the love letter of entanglement between objects and relationships. By mirroring and revealing me I am calling myself to transform and live freely as my illuminatedbeing.

Muscle testing is modeled after the same principles of the lie detector used by authorities in the pursuit of justice. When I contradict my reality my nerves will temporarily shut down and there for

the muscles become weak and can be tested to perceive my true reality in this moment. Skin resistance has a similar response and can also be ascertained.
O ring muscle test is one form of self applied muscle testing using fingers to make an o ring. See appendix of text.

'Particle' is an inevitable aspect of the quantum "wave structuring matter". All twelve of them have spin at the center of a quantum wave as the incoming wave turns around to radiate out as an outgoing quantum wave filling the universe. It lasts for an immeasurably brief period before it expands to an outgoing quantum wave. Most physicists call particles an 'appearance' of matter. To my biochemist brain they are a spinning quantum wave that matters!
The concentration and interaction of the quantum waves provide the sense of solidity in most substances.

Personalbeing describes our embodiment of our starter DNA body/brain/personality.

Promise of promising is the emergent process in the universe.

Promising is an <u>interaction</u> of our way of living.

Quantum describes the tiniest world beginning inside of the atom and expanding out to the whole universe. It includes all subatomic quantum waves and 'particles', including light, 'space' composed of 'quantum fluctuations' (probably some forms of data and/or information) as dark energy and/or dark matter, black holes, energy, and matter. The whole macro universe is quantum.

Quantumcode self = my quantum wave pattern = my abidingself (meaning: commiting to my personalself) = my eternalself (meaning: my recurring pattern throughout my living and throughout the universe recorded in quantum memory forever). These words may be used interchangeably throughout this book to help me remember the qualities of this forever companion.

Quantum 'orgasm' Is the name describing delightful physiological responses of our body when we replace the old informotion with new informotion. The form of the response depends on the substance or organ with which I am interacting. When I yawn it means I have increased my blood volume, which represents my vital emotions. It is called a 'yoyo' or yawn orgasm. It is doubled because of entanglement experience.
As I burp it means I am replacing aspects of my digestive process. We label this a bobo, a burp orgasm. And so on. This is associated with replacings. A cough might be related to lung replacement and is called a coco.
This appears to be a common observation with many forms of transformation. This term was first used by Thomas Mann in defining a sneeze which we rename 'soso' and 'snosno'. It might designate replacing something in our self image or self worth.

Random: Not having any intention or agenda. All action of the universe has no agenda and are just random action, but not chaos.

Relationshifting is a living expansive process of self-transformation as the authentic resplendentbeing that I already am. I occasionally forget this so tools are gathered here to remind me.

Resplendence is the radiant nature of my quantum wave-like being reaching out in all directions: the action of glowing. The new "flow" is glow since quantum waves "flow" instantly in all directions in the whole universe.

Smelling: Smelling is data and information (I can't smell that person)

Stem cells are "embryonic cells" that are undifferentiated and can mature into any kind of cell as directed.
Your stem self is openly new and able to do anything you desire yourself to do.
Stem self functions through nonce **memory** as a **vast abiding self eternally** when it initiates direct perception or visioning. Like **visioning it forward**... Dreams give us options to do this in our life.
You can only regenerate the body directly in two ways:

1). Do only those things you are willing to do eternally and have done to you eternally.

2). Do only those actions you are proud of doing.

This reprograms your stem self because your self is emergent from your body. So regenerating the body is the only way to regenerate the self.
Reprogram your stem cells to reprogram your stem self.

Tongue maps are love letter maps on our tongues in *Relationshifting* textbook

Traditional Chinese Medicine (TCM) is a multimillenium tested medical model of well-being based on the relationship between energy (Chi) and the material body. TCM was well developed in China and has spread to the world as a universal model.

Visioning means to see what I truly desire...The difference between visioning and visualization is that visioning is instantaneous! Words we playfully made up to describe this are visionbod or vishembody.

Uncertainty Principle, Comment on Heizenberg's, *"There is no uncertainty in quantum world. Uncertainty is in what we are knowing like the position or the momentum of a particle... not in the facts of the quantum world." (Cotting, R. B., 2018)*

'Wholgram' is a three-dimensional image formed by the interference of light beams from a laser or other coherent light source.
A photograph of an interference pattern that, when suitably illuminated, produces a three-dimensional image.
When a unique hologram photo is cut into many pieces, each piece has the whole original picture in it. This is a quantum wave principle. One piece of the wave has the pattern of the whole wave. My life mirrors this principle. One piece of my life has the same pattern as my whole life.

Glossary 2 by Roger B. Cotting (2018)

abiding: enduring (continuing) without lessening (without death as annihilation)! ...eternal!
abiding: enduring without lessening!
alternatives: different ways of achieving or accomplishing a goal!
binary: composed of two or more!
death: "to be continued!" by another, not us ...not annihilation nor is it an ending and resurrection!
emergent: appearing spontaneously from nothing! ...not to be confused with emerging in which something unfolds from within the constituents! ...nor with cause and effect! ...emergent appears, it is not caused! ...there is no causal action in emergent! ...it just appears from nothing!
direct perception: focusing on what you desire manifesting.
emmortal: having the appearance of mortal but since you will recur again and again, you are essentially immortal!
emmortalself: a self that appears mortal ...but since it recurs again and again, is actually immortal!
emprise: an enterprise or undertaking that cannot be completed in one lifetime! ...that you continue!
enrichen: same as enrich to make your self and life fuller, richer, more desirable and satisfying
enliven: make lively, act upon!
geste: a notable event or deed
life: enliven or bring to life!
lifecare: awareness and care of your abidingself and life! abiding self and life!
magination journey: direct perception of your eternal abiding self and life!
magination: direct perception!
meme: a piece of memory or a segment of memory!
mendacity: "lies," lying
MAS: memory of your abidingself!
noncetime: time varies for each occasion! This moment...
options: different ways of living-doing without a goal! ...being!
provenance: history of your abidingself!
random: without intention!!
self: the "you" that is brought to life through embodiment! ...sometimes improperly referred to as ego! ...Eternally Gifting Others... the energy aspect of you! ...self is physical! ...not spiritual!
singularity: of one! ...not composed! Has been dismissed along with theory of big bang.
soul: abiding DNA!
whole-nest: nests together; differentiated from oneness.

NOTE: I repeat some practices for two days to give myself some time to focus on it. Be gentle with myself as some may be easier for me than others. Treat the practice as an intimate date with my new forever partner every two days.
The practice changes every two days because **I will try anything twice.**

COACHING APPENDIX

Quantum Life Coaching Questionnaire
What areas do you wish to focus on?

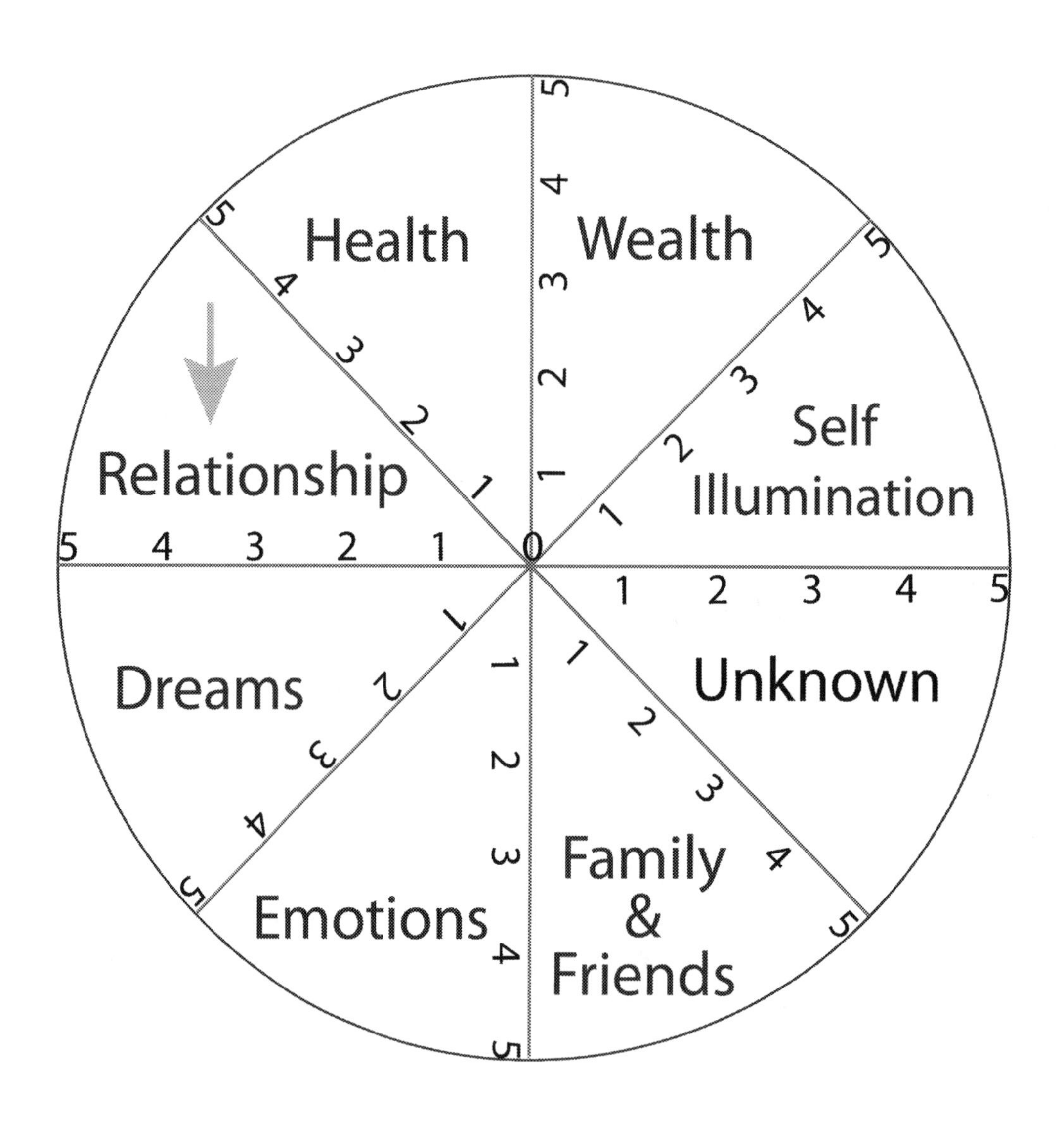

Circle the numbers you desire
5 is the most interest and 0 is the least interest

Created 2016.10.4 by Dr Angela Longo

Replacing the Eight Bottom Lines

Client's Name: ______________________ Date: ______________

	Preparing for Nonjudging	
	BATHWAVEs Part 1: Face & Embrace	Quantum Replacements Part 2: Replace with Grace
	I face and embrace that I believe that 'good and bad' exists and I am ready to replace this.	I can, I will, I do believe that good and bad do not exist.
	I face and embrace that I judge right and wrong.	I can, I will and I am living without judging. I begin to transfigure myself freely."
	Accusing	
	I face and embrace I accuse people or things.	I can, I will and I do appreciate people or things.
	I face and embrace I believe something is against me or blocking me.	The Universe can, will, and is good, benevolent, and supporting my illuminatedbeing.
	I face and embrace I feel accused by others.	I can, I will and I am valuable. I am gentle with myself.
	Judging	
	I face and embrace I judge myself and/or others.	I can, I will and I do live without judging.
	I face and embrace I feel judged by others.	I can, I will and I am valuable. I am gentle with myself.
	Blaming	
	I face and embrace I am blaming others.	I can, I will, and I do bless others for the way they are mirror-calls of my resplendency.
	I face and embrace I am blaming myself.	I can, I will, and I do understand and bless myself.
	I face and embrace I feel blamed.	I can, I will, and I do feel blessed.
	Comparing, Complaining, and Criticizing	
	I face and embrace that I am comparing, complaining, criticizing myself and/or others.	I can, I will, and I am listening (empathizing), connecting, and communicating.
	Lying	
	I face and embrace I deceive myself.	I can, I will, and I am honest and open with myself at all times.
	I face and embrace I deceive others.	I can, I will, and I am open and honest with others with discernment.
	Hiding	
	I face and embrace I am hiding from myself/others.	I can, I will, and I am visible and speak my desires.
	Denying	
	I face and embrace I have been burying things under an "unconscious" rug unknowingly.	I can, I will and I am aware of everything about me.

	Defending	
	I face and embrace I am defending myself and/or others	I can, I will and I am fine just the way I am.
	I face and embrace I am explaining myself and/or others	I can, I will and I do enjoy being the way I am without explaining.
	Justifying	
	I am justifying and proving myself to others.	I can, I will and I do as I desire. I am fine just as I am.

Client's Name: ______________________________ Date: ____________________

Metal Element - Large Intestine

		Part 1 **BATHWAVES** I Face and Embrace... in whole body	Part 2 **REPLACINGS** around my Midline with Hands, I Replace with Grace. I can, I will and I...
	Beliefs	I have to hold on to what I no longer need.	I joyfully replace the past to make space for change.
	Actions	I deplete my resources.	I use my resources appropriately.
	Thoughts	I block and sabotage my glow.	I yield to glow with living.
	Habits	I have to be in control of my own resources.	I create sustainable resources.
	Words	My words are powerless.	My words are strong and meaningful.
		I don't keep my promises.	I keep my promises.
	Attitudes	My life feels stuck.	I glow freely through living.
	Values	I believe nothing changes.	I enjoy transforming myself.
	Emotions	I feel paralysed and immobilised.	I freely dance with living.

Lung

		Part 1 **BATHWAVES** I Face and Embrace... in whole body	Part 2 **REPLACINGS** around my Midline with Hands, I Replace with Grace. I can, I will and I...
	Beliefs	I am a failure.	I attract everything I need and desire.

	Failure exists.	Everything is useful.
Actions	I have to prove myself. I am resisting living fully.	I encourage myself. I embrace living fully.
Thoughts	I reject my values.	I value myself. I live my values.
Habits	Nobody can teach me anything.	I am humble before the beauty and wonder of the universe.
Words	I am the only one who can speak the truth.	I value the words and needs of others.
Attitudes	I feel sorry for myself.	I am enthusiastic about living.
Values	I am responsible for others feelings and needs.	I honour the feelings and needs of others.
Emotions	I feel sad.	I am valuable and worthy.
	I am in grief and guilt.	I am refreshing and energizing.
	I feel insecure and unworthy, discouraged, gloomy, hopeless, unhappy, despair and intolerant.	I am significant. I am living purposeful and enlivening. I am restoring, reviving, & tolerant.

Client's Name: ______________________ Date: ________________

Element Earth
Stomach

	Part 1 **BATHWAVES** I Face and Embrace... in whole body	Part 2 **REPLACINGS** around my Midline with Hands, I Replace with Grace. I can, I will and I...
Beliefs	I worry and I am ready to replace worry.	I can, I will and I am relaxing and trusting myself and appreciating all my mirror-calls.
Actions	I am tired for the turmoil.	I can, I will and I am energizing and enlivening doing my desires.
Thoughts	I am anxious.	I can, I will and I am safe, secure, and fulfilled.
	Stomach & Pancreas: I think living is not sweet.	I can, I will and I think living is sweet.
Habits	I face and embrace in my whole body that I have an old habit of believing that things don't work out for me.	I can, I will and I am grateful that the universe nurtures me with abundant gifts.
Words	I face and embrace in my whole body that what I say doesn't make a difference.	I can, I will and I am worthy and appreciated by me.
	I face and embrace in my whole body that I do not value my own words.	I can, I will and I express myself in satisfying and fulfilling ways.

	Attitudes	I face and embrace in my whole body that I can't focus on myself.	I can, I will and believe in and focus on myself and others as I desire.
		I face and embrace in my whole body that I discontent myself.	I can, I will and... I stop discontenting myself.
	Values	I face and embrace in my whole body that I am disappointed that I don't have enough.	I can, I will and I am grateful for the abundance that I have.
		I face and embrace in my whole body that I don't value myself or others.	I can, I will and I value myself and all others in my relationships.
	Emotions	I face and embrace in my whole body that I am depressed and not truly satisfied in what I do and I am replacing that.	I can, I will and I believe in what I do and am enthusiastic about living.
		I face and embrace in my whole body that I am obsessed with things and disconnected with living and I am replacing this.	I can, I will and I focus on what I desire staying connected to my whole self.
		I face and embrace in my whole body that I am dissatisfied and feel sorry for myself.	I can, I will and I am energized, confident and appreciate all I manifest.
		I face and embrace in my whole body that I am U centered and ungrounded and I am replacing this.	I can, I will and I understand that the universe is benevolent giving me what I do in entanglement.
		I face and embrace in my whole body that I believe that this is too good to be true and I am replacing that..	I can, I will and I see my abidingbeing imaged in everything I see and do, think, feel and dream.

Client's Name: ______________________________ Date: ____________________

Organ: **Spleen/Pancreas**

		Part 1 **BATHWAVES** I Face and Embrace... in whole body	Part 2 **REPLACINGS** around my Midline with Hands, I Replace with Grace. I can, I will and I...
	Beliefs	I face and embrace in my whole body that I believe I am alienated or isolate myself from people and I am replacing that.	I can, I will and I believe I interact connecting with people in many ways like entinglement.
		I face and embrace in my whole body that I believe things happen to me and to people and I am replacing this.	I can, I will and I am happening to my living.
		I face and embrace in my whole body that I am a stranger to myself.	I can, I will and I am aware of everything about me.
		I face and embrace in my whole body that I am hard on myself and judging myself.	I can, I will and I am gentle and unconditionally loving myself.

Actions	I face and embrace in my whole body that I am withdrawn and I am replacing that.	I can, I will and I am open to reaching out and connecting.
	I face and embrace in my whole body that I react and I am replacing this.	I can, I will and I initiate and originate my living.
	I face and embrace in my whole body that I isolate and divide myself into parts.	I can, I will and I am whole and authentically me.
Thoughts	I face and embrace in my whole body that I am exhausting myself.	I can, I will and I energize myself with everything I do.
	I face and embrace in my whole body that I feel overwhelmed.	I can, I will and I focus on and dedicate myself to those things I desire.
Habits	I face and embrace in my whole body that I have a habit of needing to worry to live.	I can, I will and I relax while I do what I desire.
	I face and embrace in my whole body that my old habits are repeating.	I can, I will and I am in charge of my living and replace all habits not authentically me.
Words	I face and embrace in my whole body that my words separate, divide and are nervous.	I can, I will and I am expressing in satisfying ways as my whole resplendent illuminatedbeing.
Attitudes	I face and embrace in my whole body that I live in the environment that is obsessive and worrying and I am replacing that.	I can, I will and I live in a self sustaining supportive environment.
	I face and embrace in my whole body that I am a loner unwillingly and I am replacing this.	I can, I will and I realize that as my triunity: (my) personal, abiding and resplendent (selves), I am never alone reaching the whole universe.
	I face and embrace in my whole body that I am disconnected and I am replacing that.	I can, I will and I am in entinglement always with everything so I easily connect.
	I face and embrace in my whole body that I breathe shallowly.	I can, I will and I breath deeply in this moment.
Values	I face and embrace in my whole body that I separate my body/mind, quantum wave pattern/abidingself and resplendent illuminatedself.	I can, I will and I live my whole self without hierarchy or judging.
	I face and embrace in my whole body that I separate myself into two parts and then make up names for the two parts like masculine and feminine... and I am replacing this.	I can, I will and I am a whole functioning personalbeing living in a partnership with a whole unique abidingbeing. As a partnership we are manifesting new resplendent illuminatedbeing faster than we can imagine expanding the universe.

Emotions	I face and embrace in my whole body that I am sacrificing or martyring myself.	I can, I will and I enliven myself.
	I face and embrace in my whole body that I hurt myself.	I can, I will and I do invigorate and restore my well being.
	I face and embrace in my whole body that I feel alienated or disoriented and I am replacing that.	I can, I will and I am on track and in my orbit connecting as I desire.
	I face and embrace in my whole body that I feel stressed, worried and anxious.	I can, I will and I realize that 'stressed' spelled backwards is desserts, so I am enjoying living as a bowl full of cherries.(or box of chocolates.)
	I face and embrace in my whole body that I feel edgy and panicky.	I can, I will and I am rested, refreshed and restored.
	I face and embrace in my whole body that I feel I am obsessing and fatiguing myself.	I can, I will and I expand opening to the big picture and feel energized.

Client's Name: ______________________________ Date: ____________________

Element Water
Kidneys and Bladder

	Part 1 **BATHWAVES** I Face and Embrace... in whole body	Part 2 **REPLACINGS** around my Midline with Hands, I Replace with Grace. I can, I will and I...
Beliefs	I face and embrace that I believe that failure exists, yet I know that it is just a judgement. I am ready to replace this.	I can, I will and I am exploring many ways to do things and never give up.
	I face and embrace that I am afraid of my own power, so I don't follow my dreams.	I can, I will and I do appreciate and trust my gentle true strength to follow my dreams.
Actions	I face and embrace that I am not in charge of my living, I am ready to replace this.	I can, I will and I am in charge of my living.
	I face and embrace that I am being hard on myself, possibly out of fear.	I can, I will and I am gentle and patient with myself.
	I face and embrace that I don't listen to others opinions or needs.	I can, I will and I am open to listening to the opinions and needs of others.
	I face and embrace I hold back my energy for fear of ...(anything).	I can, I will and I am freely glowing with energy.

Thoughts	I face and embrace that I think I am not free.	I can, I will and I am free in being and doing as I desire.
	I face and embrace that I deceive myself and or others and I am ready to replace this as I desire.	I can, I will and I am impeccably honest with myself first and others as I desire.
	I face and embrace that I don't believe in myself.	I can, I will and I am believing in myself first, and in others as I desire.
	I face and embrace that I feel that I am controlled by forces or people stronger than myself.	I can, I will and I am SELF-DIRECTING my desires no matter what is happening around me.
	I face and embrace that I am indecisive.	I can, I will and I do think clearly in making decisions.
Habits	I face and embrace that I am scared to be anything.	I can, I will and I believe I am capable of fulfilling my dreams.
	I face and embrace that I belittle myself and diminish my power.	I can, I will and I am uplifting with gentle power in every interaction.
	I face and embrace that I have a habit of playing safe and small, because I am afraid to take risks.	I can, I will and I am doing new things.
	I face and embrace that I have a habit of believing I am a victim.	I can, I will and I am committed to living a fulfilling life.
Words	I face and embrace that I am afraid to speak.	My self-expression can, will and is worth hearing.
	I face and embrace that I am shy and timid.	I can, I will and I am safe and confident.
	I face and embrace that I have a hard time finding the words I desire.	I can, I will and I am a veritable thesaurus rex, or homo dictionarius. I can, I will and I am expressing myself clearly.
Attitudes	I face and embrace I have to please others and I am ready to replace that attitude.	I can, I will and I do listen and follow my self directing voice.
	I face and embrace that I am afraid to live fully.	I can, I will and I do give my whole self to living.
	I face and embrace that I distrust people and I'm ready to replace this.	I can, I will and I am discerning and trusting as I desire.
	I face and embrace that I close myself to new things and I'm ready to replace this.	I can, I will and I am open.
	I face and embrace that living is difficult and a chore and I'm ready for a new option.	Living can, will and is an illuminatedbeing delight.

	Values	I face and embrace I value others opinions more than my own and I'm ready to replace that.	I can, I will and I do value my self-directing voice.
		I face and embrace that I take better care of others than I take care of myself and I'm ready to replace that with the airplane principle.	I can, I will and I do take care of myself first, as well as taking care of others.
		I face and embrace that I am living for my past or my future and I'm ready to replace that.	I can, I will and I transform my past to my true desires in this present moment.
	Emotions	I face and embrace I am jealous or envious of myself or others. This means I do not believe I am as brilliant as OTHERS and I'm ready to replace that.	I can, I will and I am wonderful and brilliant just the way I am.
		I face and embrace that am scared, fearful and afraid, and I am ready to replace the whole shebang.	I can, I will and I am confident and secure, even with the unknown.
		I face and embrace that I am cold-hearted and I'm ready to warm things up.	I can, I will and I am generous and warm-hearted.
		I face and embrace I doubt myself and others I am ready to expand that.	I can, I will and I believe in myself and others as I desire.

Client's Name: ______________________ Date: ______________

Element Wood
Liver and Gallbladder

		Part 1 **BATHWAVES** I Face and Embrace... in whole body	Part 2 **REPLACINGS** around my Midline with Hands, I Replace with Grace. I can, I will and I...
	Essence	Face and embrace that I am discontenting myself and I would like to replace that behavior.	I can, I will and I do stop discontenting myself today.
	Beliefs	I face and embrace that I block and sabotage myself.	I can, I will and I do what benefits me
	Actions	I face and embrace that I react.	I can, I will and I do listen and initiate what I truly desire.
		I face and embrace that I put off deciding until it's too late.	I can, I will and I make timely decisions easily.

Thoughts	I face and embrace that I am resenting others or myself.	I can, I will and I do recognize that everything is beneficial.
	I face and embrace I am disappointed in others or myself.	I can, I will and I do recognize the eternalbeing in others and myself.
	I face and embrace I have trouble making choices I desire.	I can, I will and I do live my true desires with ease.
Habits	I face and embrace I am staying as I am.	I can, I will and I am pulsating my transformation throughout the universe.
	I face and embrace I am addicted and stuck to throwing in my towel to a dramatic pity party.	I can, I will and I am throwing off the towel like the veil at my belly dancing party.
Words	I face and embrace I use abrupt angry words.	I can, I will and I do hear and empathise with other people's feelings and needs.
	I face and embrace I limit myself to brain thinking.	I can, I will and I do think with the universe thanks to entinglement.
	I face and embrace I have a hard time finding words.	I can, I will and I do find words with the motion of my body and the entinglement of the universe.
Attitudes	I face and embrace I am depleting my vital emotions.	I can, I will and I am filling myself with emotions of my heart's desire.
	I face and embrace I am an irritated and or irritating.	I can, I will and I do breathe gratitude into every moment.
	I face and embrace I am bored.	I can, I will and I do stimulate my own opalescent (sparkling) radiance.
Values	I face and embrace I use am using guilt to get things done.	I can, I will and I do have the capacity and willingness to be flexible – to fearlessly leap from each sub quark of the time-continuum to the next.
	I face and embrace I think I am guilty.	I can, I will and I am innocent, refreshing and free of coercing. (forcing)
Emotions	I face and embrace I am angry.	I can, I will and I am refreshing and soothingly buoyant.
	I face and embrace I am enraging.	I can, I will and I relax and do as I desire.
	I face and embrace I am burning myself up.	I can, I will and I do fill myself with emotions of my hearts desire.
	I face and embrace that I am frustrated.	I can, I will and I do begin again new and different in each moment.
	I face and embrace that I discontent myself.	I can, I will and I do I stop discontenting myself. I can, I will and I do I live with my eternalself in one-nest.

		Hot Emotions	
		I face and embrace I am resenting.	I can, I will and I am appreciating my mirror–calls.
		I face and embrace I am bored and neglected.	I can, I will and I am creative and engaging.
		I face and embrace I am indecisive and procrastinating.	I can, I will and I am excited, decisive and just doing it.
		I face and embrace that I am depressed and nervous.	I can, I will and I am patient and self motivating in each moment.
		I face and embrace that I am disappointing/ed and self sabotaging.	I can, I will and I am cooperating with myself without judging. I can, I will and I do live as my universal illuminatedbeing authentically.

Client's Name: ______________________ Date: ______________

Element Fire
Heart

		Part 1 **BATHWAVES** I Face and Embrace... in whole body	Part 2 **REPLACINGS** around my Midline with Hands, I Replace with Grace. I can, I will and I...
	Beliefs	I face and embrace my love letter of wanting to be perfect, and believing that perfect exists.	I can, I will and I am replacing my love letter with the miracle I live my eternal authentic and real self.
		I face and embrace my love letter that bad things happen.	I can, I will and I am replacing my love letter that everything is a sacred mirror doorway of my whole self.
		I face and embrace my love letter that I'm a stranger to myself.	I can, I will and I am replacing my love letter with the miracle of being aware of my whole self.
		I face and embrace my love letter that I am judging myself harshly.	I can, I will and I am replacing my love letter with I love my aliveness unconditionally.

Actions	I face and embrace my love letter: I hide from myself.	I can, I will and I am replacing my love letter with the miracle of awakening my whole self.
	I face and embrace my love letter that I am shy or frightened in some way or another.	I can, I will and I am safe and secure no matter what's happening.
	I face and embrace my love letter that I deny myself, which creates my unconsciousness.	I can, I will and I am replacing my love letter with the miracle of appreciating how alive I am.
	I face and embrace my love letter of wanting to be perfect, that I withhold my gifts of friendship.	I can, I will and I am replacing my love letter with I listen to, acknowledge and communicate the feelings and needs of my friends.
Thoughts	I face and embrace my love letter: that I belittle myself.	I can, I will and I am replacing my love letter with having uplifting, complimentary thoughts of myself and then others.
	I face and embrace my love letter that I am avoiding thinking about myself.	I can, I will and I am replacing my love letter with I am open and aware of myself.
Habits	I face and embrace my love letter that I run away from relationships and myself.	I can, I will and I am replacing my love letter with the miracle of bringing myself anew to each relationship and myself in each moment.
	I face and embrace my love letter that I isolate myself as a loner.	I can, I will and I am replacing my love letter with the miracle that I reaching out to communicate in relationship.
	I face and embrace my love letter that I deny my entinglement mirror-calls.	I can, I will and I am replacing my love letter with appreciating the power of my entinglement.
Words	I face and embrace my love letter that I stifle and judge everything I say.	I can, I will and I am replacing my love letter with... I support resplendence with my communication.
	I face and embrace my love letter that I find it hard to talk to people.	I can, I will and I am replacing my love letter... connecting and conversing easily with people.
	I face and embrace my love letter that I can't find the words to say what I desire.	I can, I will and I am replacing my love letter with... I speak my feelings and needs honestly to others.
Attitudes	I face and embrace my love letter that I avoid myself.	I can, I will and I am replacing my love letter with... I communicate with myself and celebrate without judgement.
Values	I face and embrace my love letter that I deny the value of myself and others.	I can, I will and I am replacing my love letter... acknowledging the value of my feelings and needs and others'.

Emotions	I face and embrace my love letter that I hate what I do.	I can, I will and I am replacing my love letter with I am filled with joy and gratitude no matter what I do.

Client's Name: ______________________________ Date: ____________________

Small Intestine

	Part 1 **BATHWAVES** I Face and Embrace... in whole body	Part 2 **REPLACINGS** around my Midline with Hands, I Replace with Grace. I can, I will and I...
Beliefs	I face and embrace my love letter of believing that nothing is nourishing in my life.	I can, I will and I am replacing the old belief with perceiving that all my mirror-calls nourish me.
Actions	I face and embrace my love letter: that I block my nurturance as I live.	I can, I will and I am living that everything nurtures me.
	I face and embrace my love letter: that I interpret the actions that I encounter as hindrances or barriers.	I can, I will and I am appreciating all I encounter as a mirror-call.
Thoughts	I face and embrace my love letter: that I am distracting myself from the fullness of the moment.	I can, I will and I am replacing my love letter with my mirror-call that I am attracting everything I desire in the fullness of the moment.
Habits	I face and embrace my love letter: that I always negate my feelings.	I can, I will and I am replacing my love letter with I nourish myself with feelings I desire.
	I face and embrace my love letter that I look for affirmation from others	I can, I will and I am replacing my love letter with I affirm and celebrate my communion of self.
Words	I face and embrace my love letter that my words justify, excuse and judge me or others.	I can, I will and I am replacing my love letter with my words are emergent spontaneously and creatively.
Attitudes	I face and embrace my love letter that I have lost my desire to cook and live.	I can, I will and I am replacing my love letter with I cook and live with passion.
Values	I face and embrace my love letter that life is meaningless or draining.	I can, I will and I am replacing my love letter with everything nourishes me and gives me the meaning I choose.
Emotions	I face and embrace my love letter that I am starving for warmth and nourishment in my life.	I can, I will and I am replacing my love letter with my mirror-call that I am self- sustaining and the source of my own rejuvenation.

Client's Name: ______________________ Date: ______________

Pericardium

	Part 1 **BATHWAVES** I Face and Embrace... in whole body	Part 2 **REPLACINGS** around my Midline with Hands, I Replace with Grace. I can, I will and I...
Beliefs	I face and embrace my mirror-call that others affect me. or I face and embrace I blame and accuse others.	I can, I will and I am replacing my misunderstanding of the nature of reality with my love letter of transforming what I see to appreciate and bless everyone.
	I face and embrace my love letter of believing I am not free to be me.	I can, I will and I am replacing this with I am excited to be me as I freely desire.
Actions	I face and embrace my love letter: that I reject my mother, father, nature and the world.	I can, I will and I am replacing my love letter that the universe supports me abundantly.
	I face and embrace my love letter that I reject my mind, body, feelings and eternalself.	I can, I will and I am replacing my love letter that I am an eternal mindful being.
	I face and embrace my love letter of that I am needy for my mother, father, nature, and or the world's approval.	I can, I will and I am replacing my love letter with I nurture and guide myself without judgment.
	I face and embrace my love letter of being needy for my mind, body, feelings or being codependent with others.	I can, I will and I am replacing my love letter with my mirror-call of: I am living as a self-sustaining resplendent communion of mindful being.

	Thoughts	I face and embrace my love letter: that think I am not sexy enough.	I can, I will and I am replacing my love letter with I am a vital, sexy being.
		I face and embrace my love letter that I accuse others for my difficulties or problems.	I can, I will and I am replacing my love letter with I appreciate everything is a sacred doorway for my illuminated being.
		I face and embrace my love letter that I blame others for my difficulties or problems.	I can, I will and I am replacing my love letter with I bless others as sacred doorways for my illuminated being.
		I face and embrace my love letter that I complain about others or myself and discontent myself.	I can, I will and I am replacing my love letter I connect and communicate as my illuminated being without discontenting myself.
		I face and embrace my love letter that difficulties and problems exist.	I can, I will and I am replacing my love letter with everything is my mirror-call for my universal illumination.
		I face and embrace my love letter that I think that things are opposing me in the universe.	I can, I will and I am replacing my love letter with everything supports me as I support myself.
		I face and embrace my love letter that the thought of failure motivates me.	I can, I will and I am replacing my love letter with no matter what occurs I am successful, sharing and inspiring my new self.
	Habits	I face and embrace my love letter: that I am guilty.	I can, I will and I am replacing my love letter that I am innocent.
		I face and embrace my love letter that I am resentful as I live.	I can, I will and I am replacing my love letter with I desire to communicate a wonderful life to everyone I meet.
		I face and embrace my love letter that I am overbearing in what I do.	I can, I will and I am replacing my love letter I listen and empathize with the feelings and needs of others.
		I face and embrace my love letter that I use force, control and power to manipulate to get things done.	I can, I will and I am replacing my love letter with as I transform myself and watch others and things transform around me.
	Words	I face and embrace my love letter that my words are alienating and isolating.	I can, I will and I am replacing my love letter that my words express the union of communion of the quantum way.
		I face and embrace my love letter that others words trigger me.	I can, I will and I am replacing my love letter that I unconditionally listen to others ideas, feelings and needs.

	Attitudes	I face and embrace my love letter that my negative attitudes block my clarity and strength of self.	I can, I will and I am replacing my love letter with my true strength is my personal change.
		I face and embrace my love letter that I doubt my true strength and wisdom.	I can, I will and I am replacing my love letter with I have full faith in my personal change.
	Values	I face and embrace my love letter that there is no free will.	I can, I will and I am replacing my love letter with I value my freewill to keep or recycle whatever I desire.
		I face and embrace my love letter that the universe is out to get me.	I can, I will and I am replacing my love letter with the universe supports me as I support myself.
	Emotions	I face and embrace my love letter that fear of failure paralyzes me.	I can, I will and I am replacing my love letter with I understand that there are no mistakes or failures. I can, I will and I do never give up.
		I face and embrace my love letter that I feel disoriented and out of my orbit.	I can, I will and I am replacing my love letter with I am fired up with enthusiasm.

Client's Name: ______________________ Date: ______________

Triple Heater

		Part 1 **BATHWAVES** I Face and Embrace... in whole body	Part 2 **REPLACINGS** around my Midline with Hands, I Replace with Grace. I can, I will and I...
	Beliefs	I face and embrace my love letter of believing that my world is not in harmony with me.	I can, I will and I am replacing the past with my mirror-calls of living are in harmony with me, for my transformation.
		I face and embrace my love letter: that I believe that I should judge.	I can, I will and I am replacing my love letterl that good and bad do not exist...I am willing to reflect on the mirror-call of what is really going on in me.
	Actions	I face and embrace my love letter: that I am disoriented and lost.	I can, I will and I am living my eternalself in each moment.
	Thoughts	I face and embrace my love letter: that my life is not in harmony with me.	I can, I will and I am replacing my love letter with my mirror-call that everything is in "entinglement" with me therefore it's always in harmony with me.
	Habits	I face and embrace my love letter: that I am always trying to change the world.	I can, I will and I am replacing my love letter with my mirror-call that as I change myself I am changing the world.

Words	I face and embrace my love letter that my words are cold and judgmental.	I can, I will and I am replacing my love letter with...my words are warm and empathic.
Attitudes	I face and embrace my love letter that I am depressed and despairing at this disharmony.	I can, I will and I am replacing my love letter with my mirror-call that I am full of elation as I dance with the glow of living.
Values	I face and embrace my love letter that my feminine side does not value my masculine side.	I can, I will and I am replacing my love letter with that my feminine side (feeling, poetic, being side) values my masculine side (linear, logical thinking, organized side).
Emotions	I face and embrace that I hate the disharmony of relationships.	I can, I will and I do delight in stimulating relationships.
	I face and embrace my love letter that I close my energy down in relationships.	I can, I will and I am energizing myself in relationship.

Bibliography and Reading List

Al-Khalili, J. (2017) Quantum Mechanics (A Ladybird Expert Book)

Al-Khalili, J. (2013) Quantum Life: How Physics Can Revolutionize Biology Youtube Ri.

Asimov. I. (1972). *The gods themselves.* New York, NY: Bantam Books.

Baerbel. (2006). *DNA can be influenced and reprogrammed by words and frequencies: Russian DNA discoveries.* Retrieved July 21, 2012, from http://www.soulsofdistortion.nl/dna1.html.

Becker, R. (1990). *Cross currents: the promise of electromedicine, the perils of electropollution.* New York, NY: J P Tarcher.

Beinfield, H. (1991). *Between heaven and earth: a guide to chinese medicine.* New York, NY: Ballantine Books.

Benveniste, J. (1988). Water Memory. Nature, 333:816-818.

Bethell, T. (2009). *Questioning Einstein: is relativity necessary?* Pueblo West, CO: Vales Lake Publishing.

Bonn, E. (2009). Turbulent contextualism: bearing complexity towards change. *International Journal of Psychoanalytic Self-Psychology.* Volume 5, Issue 1, pp 1-18.

Braden, G. (2008). *The spontaneous healing of belief: shattering the paradigm of false limits.* USA: Hay House.

Brailsford, B. (2004). *Song of the old tides.* Christchurch, NZ: Stoneprint Press.

Briggs, J. & Peat, F. (1989). *Turbulent mirror.* New York, NY: Harper and Row Publishers.

Capra, F. (1996). *The web of life: a new scientific understanding of living systems.* New York, NY: Random House.

Cerney, J. (1974/1999). *Acupuncture without needles.* Paramus, NJ: Prentiss Hall.

Chia, M. (2006). *Chi self-massage: the Taoist way of rejuventation.* Rochester, VT: Destiny Books.

Childre, D. (2004). *From chaos to coherence: the power to change performance.* Boulder Creek, CA: HeartMath LLC.

Chopra, D. (1989, 2015). *Quantum healing: exploring the frontiers of mind/body medicine.* New York, NY: Bantam Books.

Coelho, P. (1993). *The alchemist.* New York, NY: HarperCollins Publishers.

Colton, A. (1973). *Watch your dreams: a master key and reference book for all initiates of the soul, the mind and the heart.* Glendale, CA: The Ann Ree Colton Foundation.

Connelly, D. (1979*). Traditional acupuncture: the law of the five elements.* Columbia, MD: Traditional Acupuncture Inc.

Cotting, R. B. (2018). Book of Origins. e-book Kindle.

Cotting, R. B. (2012). Emergentmiracles.net Pericope 14A.

Croca, J.R. (2003). *Towards a nonlinear quantum physics:* River Edge, NJ: World Scientific Publishing, Co.

Dispenza, J. (2008). Evolve your brain. USA: Health Communications.

Doidge, N. (2007). *The brain that changes itself: stories of personal triumph from the frontiers of brain science.* New York, NY: Penguin Books.

Emoto, M. (2005). *The hidden messages in water.* New York, NY: Atria Books.

Emoto, M. (2005). *The true power of water.* New York, NY: Atria Books.

Feinstein, D. (2003). *Energy psychology interactive self-help guide.* Ashland, OR: Innersource.

Feldenfrais, M. (1949,2005). *Body and mature behavior: a study of anxiety, sex, gravitation, and learning.* Berkeley, CA: Frog, Ltd.

Feldenkrais, M. (1972). *Awareness through movement: health exercises for personal growth.* New York, NY: HarperCollins.

Feng, G. and English, J (1972). *Tao te ching.* New York, NY: Random House.

Flaws, B. (1995). *The secret of chinese pulse diagnosis.* Boulder, CO: Blue Poppy Press.

Fosar, G., Bludorf, F. (2001). *Vernetzte Intelligenz.* Aachen, Germany: Omega Verlag.

Frankl, V. (1946/1997*). Man's search for meaning.* New York, NY: Pocket Books.

Garfield, P. (1981). Creative dreaming NY, Ballantine Books,

Garfield, P. (2001).*The universal dream key: the twelve most common dream themes around the world.* New York, NY: HarperCollins.

Geldard, R. (2007). Parmenides and the way of truth; translation and commentary. Rhinebeck, NY: Monkfish Book Publishing Company.

Gennaro, L. (1980). *Kirlian photography: research and prospects.* UK: East-West Publications Ltd.

Gillis, L., Suler, J. (2001). *Dream secrets: unlocking the mystery of your dreams.* Lincolnwood, IL: Publications Internationals, Ltd.

Gleick, J. (1987). *Chaos: making a new science.* New York, NY: Penguin Books.

Goertzel, B. (2006). *The hidden pattern: a patternist philosophy of mind.* Boca Raton, FL: BrownWalker Press.

Golomb, E. (1992). *Trapped in the mirror.* New York, NY: William Morrow and Company.

Goswami, A. (1993). *The self-aware universe: how consciousness creates the material world.* New York, NY: J P Tarcher.

Greene, B. (2004). *The fabric of the cosmos: space, time and the texture of reality.* New York, NY: Vintage Books.

Hannaford, C. (2002). *Awakening the child heart: handbook for global parenting.* Captain cook, HI: Jamilla Nur Publishing.

Hannaford, C. (2010). *Playing in the unified field: raising and becoming conscious, creative human beings.* Salt Lake City, UT: Great River Books.

Haselhurst, G. (1997-2018). www.spaceandmotion.com

Hawkings, S. (2012). The grand design. USA: Bantam books.

Hay, L. (1991). *The power is within you.* USA: Hay House.

Hay, L. (1999). *You can heal your life.* USA: Hay House.

Hickok, G. (2014). The myth of mirror neurons. USA: W. W. Norton and Co.

Hoff, B. (1982). *The tao of Pooh.* New York, NY: Penguin Books.

Hoff, B. (1992). *The te of piglet.* New York, NY: Penguin.

Iovine, J. (2000). *Kirlian photography: a hands-on guide.* USA: Images Publishing.

Jantsch, E. (1980). *Design for evolution: self-organization and planning in the life of human systems.* Elmsford, NY: Pergamon Press.

Johnjol, M., and Al Kahlili, J. (2015) Life on the Edge: The Coming of Age of Quantum Biology. USA: Broadway Books, Penguin Random House.

Kako, M. (2011). Physics of the future. USA: Doubleday.

Kelly, R. (2006). *The human antenna: reading the language of the universe in the songs of our cells.* Santa Rosa, CA: Energy Psychology Press.

Lane, E. (1975). *Electrophotography.* San Francisco, CA: And/Or Press.

Laskow, L. (1992). *Healing with love: a physician's breakthrough mind/body guide for healing yourself and others; the art of holoenergetic healing.* Mill Valley, CA: Wholeness Press.

Leu, L. (2003). *Nonviolent communication companion workbook: a practical guide for individual, group, or classroom study.*, CA: PuddleDancer Press.

Liangyue, D. (1987). *Chinese acupuncture and moxibustion.* Beijing, China: Foreign Languages Press.

Lipton, B. (2008). *Biology of belief: unleashing the power of unconsciousness, matter and miracles.* USA: Hay House

Lipton, B. (2009). *Spontaneous evolution: our positive future (and a way to get there from here).* USA: Hay House.

Long, M.F. (1948). *The secret science behind miracles.* Camarillo, CA: DeVorss & Company.

Longo, A. (2015). Angelalongo.com Quantum Wave of Living. 1-6. Youtubes

Longo, A. & Penhoet, E. (1974). *Nerve Growth Factors in Rat Glioma Cells.* Proceedings of the National Academy of Sciences Vol. 71, No. 6, pp. 2347-2349, June 1974.

Longo, A. & Penhoet, E. (1974). *Nerve Growth Factor Synthesis in Rat Glioma Cells.* Federal Proceedings, 33, (5) 1409.

Longo, A. (2012). *Relationshifting; Tools For Living Quantum Resplendence* USA: iUniverse.

Longo, A. (1978). *Synthesis of Nerve Growth Factor in Rat Glioma Cells.* Developmental Biology 65, 260-270, 1978.

Mahoney, M. (1966). *The meaning of dreams and dreaming: the Jungian viewpoint.* Secaucus. NJ: Citadel Press.

Mann, F. (1962/1971). *Acupuncture: the ancient chinese art of healing and how it works scientifically.* New York, NY: Random House.

Marcher, L. Fich, S. (2010). *Body encyclopedia: a guide to thepsychological functions of the muscular system.* Berkeley, CA: Atlantic Books

McTaggart, L. (2002). *The field.* New York, NY. HarperCollins.

Martini, F. (1993). *The meaning of your dreams.* Baltimore: MD: Ottenheimer Publishers.

Meijer, D. (2017). *NeuroQuantology,* September issue.

Moalem, S. (2007). *Survival of the sickest: a medical maverick discovers why we need disease.* New York, NY: HarperCollins Publishers.

Montagnier, L. (2016) Water Memory (2014 documentary about research on water) Youtube.

Montagu, A. (1971). *Touching: the human significance of the skin.* New York and London: Columbia University Press.

Moss, T. (1980). *The body electric: a personal journey into the mysteries of parapsychology and kirilian photography.* Los Angeles, CA: J P Tarcher.

Nuernberger, P. (1981). *Freedom from stress: a holistic approach.*

Honesdale, PA: Himalayan International Institute of yoga Science and Philosophy Publishers.

O'Hanlon, W. (1995). *Stop blaming, start loving: a solution-oriented approach to improving your relationship.* New York, NY: W. W. Norton & Company.

Pearce, J. (1977). *Magical Child.* New York, NY: Bantam Books.

Pearce, J. (2002). *The biology of transcendence: a blueprint of the human spirit.* Rochester, VT: Park Street Press.

Pearsal, P. (1998). *The heart's code.* New York, NY. Broadway Books.

Peck, M. (1987). *The different drum.* New York, NY: Touchstone.

Pelletier, K. (1977). *Mind as healer, mind as slayer.* New York, NY: Dell Publishing Co.

Pert, C. (1997). *Molecules of emotion: why you feel the way you feel.* New York, NY: Scribner.

Quintin, J. (2011). *The unity of geometry.* DVD: NZ: self-published.

Radin, D. (2006). *Entangled minds: extrasensory experiences in a quantum reality.* New York, NY: Pocket Books.

Reps, P. (1968). Square Sun, Square Moon. *Gold and fish signatures.* Rutland, VT: Charles Tuttle Company.

Roth, G. (1998). *Maps to ecstacy: the healing power of movement.* Novato, CA: New World Library.

Satir, V. (1976). *Making contact.* Millbrae, CA: Celestial Arts.

Satprem. (1982). *The mind of the cells.* New York, NY: Institute for Evolutionary Research.

Scheele, P. (1996). *Natural brilliance: overcoming any challenge – at will.* Wayzata, MN: Learning Strategies Corp.

Scherer, K. (1990) KarlScherer.com (author of fractal art).

Schonberger, M. (1992). *I Ching and the genetic code: the hidden key to life.* Santa Fe, NM: Aurora Press. New York, NY: Berkley Books.

Siegel, B. (1986). *Love, medicine and miracles.* New York, NY: Harper & Row.

Siegel, D. (2011). *Mindsight.* Los Angeles, CA: Mindsight Institute. Roads.

Smith, C. (2000). *Descriptionary: inpicturing, outpicturing.* USA: Wisdom of the body.net.

Starwynn, D. (2002). *Microcurrent electro-acupuncture: bio-electric principles, evaluation and treatment.* Phoenix, AZ: Dessert Heart Press.

Steiner, R. (1959). *Cosmic Memory: Atlantis and Lemuria.* Blauvelt, NY: Rudolf Steiner Publications.

Stone, R. (1989). *The secret life of your cells.* West Chester, PA: Whitford Press.

Tanner, W. (1988). *The mystical, magical marvelous world of dreams.* Tahlequah, OK: Sparrow Hawk Press.

Tebo, B. (1993). *Free to be me.* New York, NY: Bantam Books.

Tegmark, M. (2015). Our mathematical universe: my quest for the ultimate nature of the universe. USA: Vintage Books.

Thakar, V. (1974). *Five talks by Vimala Thakar on the general subject of the nature of human consciousness, its challenges and problems, and its transformation through meditation.* Claremont,

CA: The Blaisdell Institute.

Tiller, W. (1997). *Science and human transformation: subtle energies, intentionality and consciousness.* Walnut Creek, CA: Pavior Publishing.

Tiller, W. (2001). *Conscious acts of creation: the emergence of a new physics.* Walnut Creek, CA: Pavior Publishing.

Tiller, W. (2005). *Some science adventures with real magic.* Walnut Creek, CA: Pavior Publishing.

Tong, D. (2017) Quantum Fields: The Real Building Blocks of the Universe. YouTube.

Vitale, J. & Behrend, G. (2004). *How to attain your desires by letting your subconscious mind work for you.* Newport News, VA: Morgan-James Publishing.

Warren, R. (2002). *The purpose driven life: what on earth am I here for?* Grand Rapids, MI: Zondervan.

Waters, F. (1963). *Book of the Hopi: the first revelation of the Hopi's historical and religious world-view of life.* New York, NY: Ballantine Books.

Weiss, K. (1984). *Women's health care: a guide to alternatives.* USA: Reston Publishing Company.

Wheeler, J.A. (2004*). Science and ultimate reality; quantum theory, cosmology and complexity.* Cambridge, UK: Cambridge University Press.

Whorf, B. (1956) Language, thought, and reality. USA: M.I.T.

Wilhelm, R. (1967). *I Ching.* Princeton, NJ: Princeton University Press.

Williams, R. (1956). Biochemical individuality. New Cannan, Conn. USA: Keats Publishing.

Wolff, M., (2008). *Schrodinger's Universe: Einstein, Waves and the Origin of the Natural Laws.* Parker, CO: Outskirts Press.

Woolf, V. (1990). *Holodynamics: how to develop and manage your personal power.* Tucson, AZ: Harbinger House.

Xuemei, L. and Jingyi, Z. (1993). *Acupuncture patterns and practice.* Seattle, WA: Eastland Press.

Yap, S. & Hiew, C. (2002). *Energy medicine in CFQ healing.* USA: iUniverse.

Zukav, G. (1989). *The seat of the soul.* New York, NY: Fireside.

APPENDIX NOTES ON PHYSICS: NERD ALERT! OPTIONAL QUANTUM WAVE THEORY FOR THOSE WHO DESIRE

QUANTUM THINKING VERSUS ONLY BRAIN THINKING:

ENTANGLEMENT PERCEIVED ON HUMAN SCALE

I have been gathering scientific theorizing and evidence in the field of neuroscientific studies, in which I got a Ph.D from UCB. Quantum Biology has already demonstrated the presence of entanglement in photosynthesis and in the eyes of the robin. Kahlil, J. (2013)

"Dr. Dirk K.F. Meijer, a professor at the University of Groningen in the Netherlands, hypothesizes that consciousness resides in a field surrounding the brain. This field is in another dimension. **It shares information with the brain through quantum entanglement, among other methods."** Meijer, D. (2017)

"Information comes together and interacts in the brain more quickly than can be explained by our current understanding of neural transmissions in the brain. **It thus seems the mind is more than just neurons firing in the brain."** Meijer, D. (2017)

"**Rather, *quantum wave resonance* is a more likely mechanism of extremely rapid information processing in the brain.** This means, *instead of signals being sent between neurons in the brain,* ***a wave pattern that encompasses all neurons, as well as the mental field, transmits the information instantaneously."*** Meijer, D. (2017)

"Picture a vibration wave going up and down in a consistent pattern and running all through your brain and even outside of it. That pattern communicates information that can be understood by vibratory receptors in your brain. All of this is happening in a dimension and at a microscopic level not directly perceptible through conventional scientific instrumentation at our disposal today, yet can be inferred through physical and mathematical modeling." Meijer, D. NeuroQuantology, (September 2017)

"A groundbreaking quantum experiment recently confirmed the reality of "spooky action-at-a-distance" — the bizarre phenomenon that Einstein hated — in which linked particles seemingly communicate faster than the speed of light." Mitchell, M. (2018)

"And all it took was 12 teams of physicists in 10 countries, more than 100,000 volunteer gamers and over 97 million data units — all of which were randomly generated by hand.
The volunteers operated from locations around the world, playing an online video games on Nov. 30, 2016, that produced millions of bits, or "binary digits" — the smallest unit of computer data." Mitchell, M. (2018)

"Study co-author Morgan Mitchell, a professor of quantum optics at the Institute of Photonic Sciences in Barcelona, Spain, told Live Science in an email.
"We showed that Einstein's world-view of local realism, in which things have properties whether or not you observe them, and no influence travels faster than light, cannot be true — at least one of those things must be false," Mitchell said.
This introduces the likelihood of two mind-bending scenarios: **Either our observations of the world actually change it, or particles are communicating with each other in some manner that we can't see or influence.**
"Or possibly both," Mitchell added." Mitchell, M. (2018)

Both seem possible as interactive entanglement. When I live my quantum wave patternself with awareness I am potentially free of impact by entanglement as I desire.

"WAVES ARE PRODUCED BY ENERGY EXCHANGES"*

*Wolff, M. (2008)

QUANTUM WAVES ARE THE UNIVERSE AND SPACE

Physicist Wheeler said,
"Matter tells space what it is. Space tells matter how to behave."

Since all matter is quantum waves then it follows:

"Quantum waves define space. (quantum waves of dark light). Space informs quantum waves (dark light) how to behave (through random entanglement)." Longo, A. (2018)

Space is random motion. (energy is data motion)
Space randomly manifests informotion. (dark data coalescing)

"Quantum waves tell energy (space) what it is. Energy (Space) informs quantum waves how to behave."
See figure 1.2A data and informotion to understand the next two thoughts:
Dark energy = "Dark data"*
Dark matter = "Dark informotion" Longo, A. (2018)

Sharing Our In and Out Waves

Other matter's spherical out-waves form our matters in-waves. This is called Huygen's principle. Four important points:

1). In reality there are about 10 to the 80th other waves center particles whose spherical out-waves form into our spherical in-waves.

2). They are obviously not all the same distance away, but distributed throughout the space of our finite spherical observable universe (Hubble Sphere) within infinite eternal space.

3). These other wave-center "particles" around us are also formed from the matter waves around them, and this process extends to infinity.

4). This is difficult to diagram because there is no beginning or ending to this process. In reality the waves are continually flowing out from other matter around us. This means that the system is perpetual, the in and out waves are always being shared between electron wave centers wherever you are in infinite space. Like entinglement.

So this is a very simplistic explanation of what is really going. Electrons are very large complex wave structures of the universe (this is true for humans too). Hazelhurst, G. (2012)

See Figure 16.2. Wolff, M. (2008)

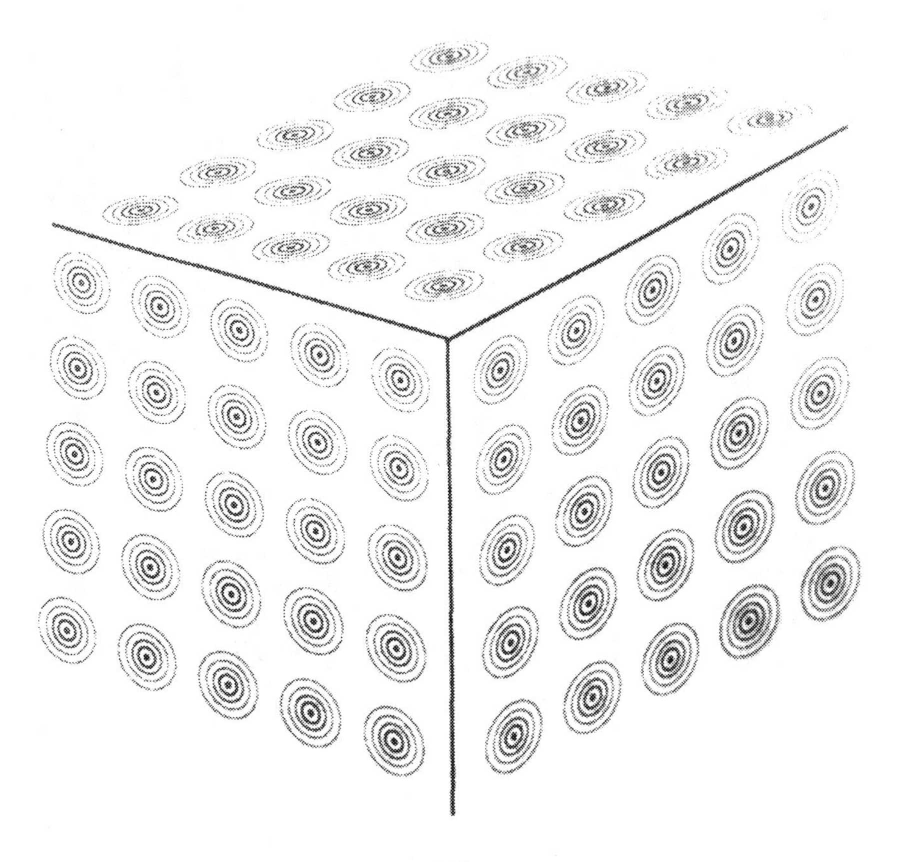

"Figure Atoms in a crystal. (Like a diamond)
The waves of the electrons in the crystal array produce standing waves along the planes of symmetry. There is no solid structure, substance or material in the crystal. It is the waves traveling in the space medium that produce the array dimensions, and it is the immense energy density of the space medium that gives it physical strength and rigidity.
I. Origin of Solid Matter.
The solid structures of everyday life are held together by wave structures that occur because of the Minimum Amplitude Principle that permits only solids with minimum total wave amplitudes. Solidity and rigidity are properties of the waves –there is no solid substance. For example atoms in say a diamond, are held in a rigid lattice by standing electron waves in the space medium traveling throughout the carbon lattice." Wolff, M. (2008) p64

STRENGTH OF A SOLID LIKE A CRYSTAL

"The strength of solids depends on the forces that produce the waves; these depend on the energy density of space; that is. the wave medium. The minimum energy density of space can be estimated as the density of nuclear matter. It is more than 1046 MeV per cubic centimeter, astoundingly large. We have no sensation of its presence despite its existence all around us." Wolff, M. (2008) p64

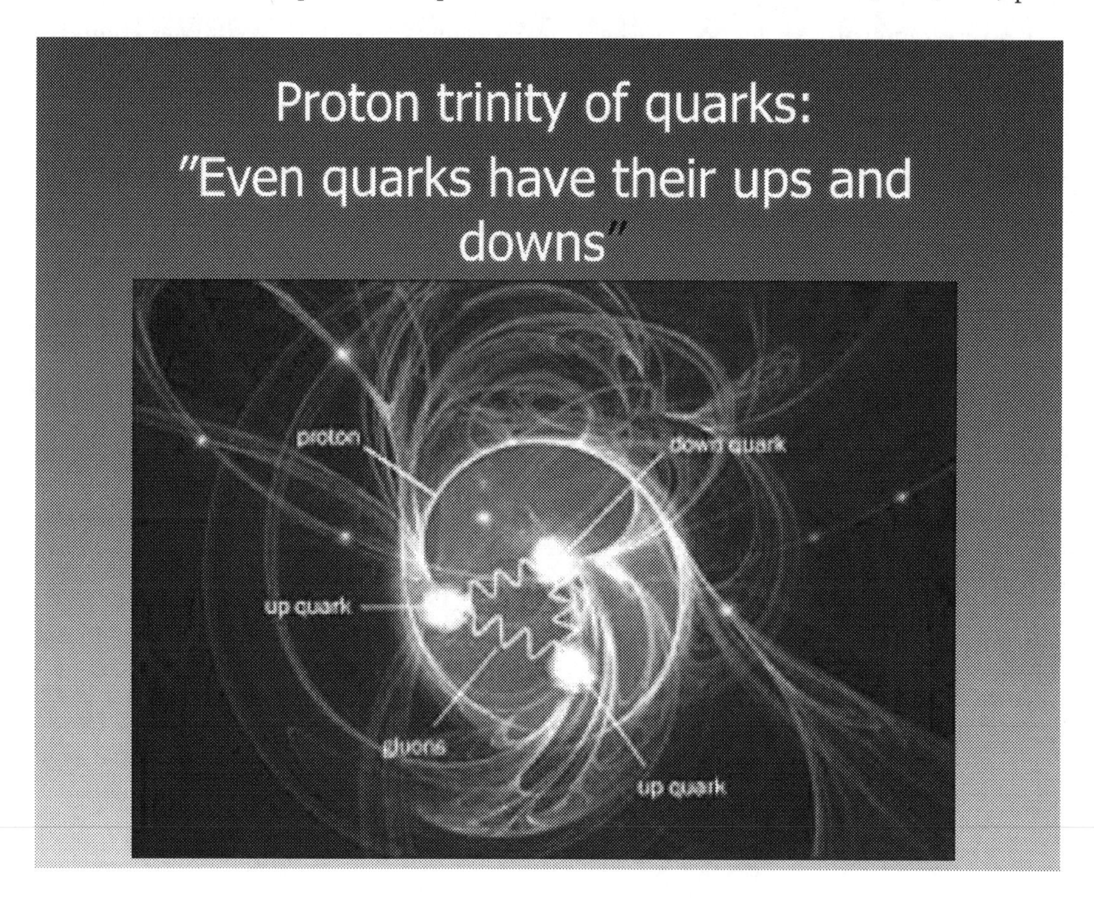

About the Author

Dr. Angela Longo received her Doctorate in Biochemistry from UC Berkeley published original discoveries in the field of neurochemistry of glia/nerve interaction initiating the importance of glia in the brain and introduced the importance of small peptides in the nervous system. Her traditional training in Chinese Medicine was from Dr. Lam Kong, a teacher with a legacy of ten generations in Chinese Medicine.

With 46 years experience as a Licensed Practitioner of TCM, a professor of interdisciplinary holistic science at San Francisco State University, founding a college of Traditional Chinese Medicine in Hawaii with 46 years as a teacher, she brings a wealth of knowledge to her practice of Traditional Chinese Medicine and Quantum Life Coaching.

Dr. Longo teaches Quantum Wave Living, Relationshifiting and Resplendency internationally and trains one-on-one Quantum Wave Coaching. She lives both in Kamuela, Hawaii and Chiang Mai, Thailand.

Contact her at angelalongo1@gmail.com .

Visit angelalongo.com

Printed in the United States
By Bookmasters